Muscles
in Action

AN APPROACH TO MANUAL MUSCLE TESTING

Joan H. Cole BPT MS MEd

Senior Lecturer, School of Physiotherapy, Curtin University of Technology, Perth, Australia

Anne L. Furness MAppSci AssDipPhysiotherapy, GradDipHlthSc

Lecturer, School of Physiotherapy, Curtin University of Technology, Perth, Australia

Lance T. Twomey BApp Sci (Phty) BSc (Hons) Phd

Professor and Head, School of Physiotherapy, Curtin University of Technology, Perth, Australia

Photographs by
David Watkins

Audiovisual Technician, School of Occupational Therapy, Curtin University of Technology, Perth, Australia

Churchill Livingstone

MELBOURNE EDINBURGH LONDON AND NEW YORK 1988

CHURCHILL LIVINGSTONE
Medical Division of Longman Group UK Limited

Distributed in Australia by Longman Cheshire Pty Limited, Longman
House, Kings Gardens, 95 Coventry Street, South Melbourne 3205,
and by associated companies, branches and representatives throughout
the world.

First published 1988

ISBN 0-443-03613-6 ✓

British Library Cataloguing in Publication Data
Cole, Joan H.
    Muscles in action : an approach to manual muscle testing.
    1. Muscles
    I. Title   II. Furness, Anne L.
    III. Twomey, Lance T.
    612'.74   QP321

Library of Congress Cataloging in Publication Data
Cole, Joan H.
    Muscles in action.
    Bibliography: p.
    Includes index.
    1. Muscles — Examination. 2. Neuromuscular diseases —
    Diagnosis. I. Furness, Anne L.   II. Twomey, Lance T.   III. Title.
    [DNLM: 1. Muscles — physiology. WE 500 C689m]
    RC925.7.C65   1988   616.7'4075   87–14629

Produced by Longman Singapore Publishers (Pte) Ltd.
Printed in Singapore

Preface

This text is designed to offer students of a variety of disciplines a structured introduction to the evaluation of strength and function of the muscles of the human body. It is also intended as an easy reference for the clinician in practical situations.

The book is presented in two parts. The initial chapters address the fundamentals of muscle testing, and the problems associated with it. This introduction is intended to remind the user that this method of evaluation has specific limitations which must be understood if testing is to provide a reliable measure of performance. The major component of the text lists the individual muscles, together with their attachments, nerve supply, surface markings, actions and the position of maximum extensibility. The test position for most muscles of the body against the resistance of gravity and with gravity eliminated are provided. The principal testing position is demonstrated pictorially in each case and, where a major modification of the basic position is required, this has been presented also.

Commonly, manual muscle testing is directed towards those individuals with functional disabilities of the musculoskeletal system. Reference to a variety of muscles which are not primarily involved in damage to the peripheral nervous system is omitted from muscle testing manuals, as a general rule. The present approach takes the view that muscle weakness is a frequent finding in all instances of neuromuscular dysfunction. Therefore abilities such as eye movements, swallowing and other essential functions have been included in an endeavour to broaden the scope of application of the evaluation of muscle.

Every effort has been made to identify those clinical situations which frequently make muscle testing more difficult. Some of the pitfalls likely to interfere with the evaluations have been described, and a commonsense approach to testing has been suggested. In addition, to assist the beginning practitioner, special attention has been directed towards the efficient application of muscle testing in dealing with the most commonly encountered pathologies. The sequencing of the testing procedure in lying, sitting and standing is presented in a tabular form as are the muscles which need to be evaluated in each of the major peripheral nerve injuries and in lesions of the spinal cord at a variety of levels.

The objective of the authors in compiling this manual has been to provide a degree of consistency both in organization and execution not found in other books which describe the process of muscle testing. It is hoped that this approach may increase the reliability of manual muscle testing by encouraging all who utilize the system to record reproducible results. Perhaps rehabilitation will be enhanced as a science if sufficient numbers of health professionals consider their methods of evaluation in this manner and strive for consistency in their techniques.

Joan Cole
Anne Furness
Lance Twomey

Perth, 1988

Acknowledgements

The authors wish to pay special tribute to those who have made a major contribution to this book. David Watkins, the photographer, who contributed many hours and much originality. The models who suffered cheerfully, through innumerable photographic sessions and the typists, Faye Gooch, Jenny Cook, Megan Carwardine and Joan Reyland, who remained calm through all the moments of stress, all deserve our praise and appreciation. The contribution of Peter, Paul and Meg, who kept the dinner warm so often, is without equal. Perhaps, we can return the service, in part, by gratefully acknowledging these contributions.

Contents

PART 1
Fundamentals of
muscle testing

1 **Introduction** / *3*
    Types of muscle contraction / *3*
    Limitations of manual muscle testing / *4*
    Standardizing the procedure for manual muscle testing / *5*
    Concepts in test construction / *5*
    Grading system / *7*
    Accessory descriptors / *8*
    Sequence of presentation / *9*
    Additional elements / *10*

2 **Clinical guidelines in manual muscle testing** / *11*
    Optimal preparation for testing / *11*
    Legitimate modifications / *12*
    Special problems / *15*
    Aids to testing and diagnosis / *15*
    Muscle grading charts / *17*
    Figures / *18*
    Tables / *21*

PART 2
Muscle tests

3 **Evaluating the muscles of the upper extremity** / *30*
    Muscles tested at the scapula / *30*
    Muscles tested at the shoulder joint / *38*
    Muscles tested at the elbow joint / *48*
    Muscles tested at the superior and inferior radio-ulnar joints / *53*
    Muscles tested at the wrist joint / *56*
    Muscles tested at the joints of the fingers / *61*
    Muscles tested at the joints of the thumb / *73*

4 **Evaluating the muscles of the lower extremity** / *80*
    Muscles tested at the hip joint / *80*
    Muscles tested at the knee joint / *91*
    Muscles tested at the ankle joint / *96*
    Muscles tested at the joints of the toes / *103*
    Weight bearing tests of the lower extremity / *113*

5 **Evaluating the muscles of the neck and trunk** / *118*
    Muscles which move the joints of the cervical spine / *118*
    Muscles which move the joints of the thoracic, sacral, and lumbar spine / *122*

6 **Evaluating the muscles of the face** / *129*
    Muscles of mastication / *129*
    Muscles of facial expression / *129*
    Muscles of the eye / *140*

7 **Evaluating the muscles of the essential functions** / *143*
    Evaluating the muscles of swallowing / *143*
    Evaluating the muscles of respiration / *150*
    Evaluating the muscles of micturition and defaecation / *152*

Bibliography / *155*

Appendices / *156*
    Muscle Grading Chart 1
        Trunk, Upper Limb, Lower Limb / *156*
    Muscle Grading Chart 2
        Face and Essential Functions / *163*

Index / *167*

Fundamentals of muscle testing

Movement is an essential part of being. It is an expression of the various mechanisms of the neural control system and, in humans, a manifestation of the thought processes in which the individual is engaged. Without movement there are remarkably few behaviours which can be exhibited (Wolff 1981).

Apart from some glandular reactions such as the production of sweat or tears, most species are dependent on muscle tissue for interaction with the environment. Thus muscles are the vital transducers of the body, transforming complex sequences of chemical and electrical events into mechanical actions which result in movement.

Since movement and the mechanism of its production are such a central component of the system by which human activity is displayed, it is to be expected that evaluating muscle capacity is of major importance to physiotherapists, medical practitioners, and others in health and related professions who are concerned with habilitation and rehabilitation. This text describes the way in which muscle strength is most usually evaluated. The technique is known as manual muscle testing and other authors (Daniels & Worthingham 1986, Kendall & McCreary 1983, Pact et al 1984) have outlined its origins. Manual muscle testing is designed to evaluate rapidly developed tension in individual muscles or groups of muscles. The technique requires the application of increasing amounts of resistance to produce a concentric contraction of the muscle to be tested, until a maximum contraction is reached. The increasing resistance is afforded initially by the weight of the segment to be moved. To this is added gravity and then manual resistance.

Manual muscle testing has become an integral part of the evaluation of neuromotor performance in infants who have spina bifida, in children with muscular dystrophy, in young people with spinal cord injuries and peripheral nerve lesions, and in a host of other disorders in all age groups where the capacity for normal behavioural interaction with the environment has been diminished. The description of the technique presented in this text is designed to form a part of the assessment of those individuals in whom motor patterns have been interrupted or disrupted whether by prenatal influences, injury, disease or lifestyle. In order to appreciate the limitations of manual muscle testing in the assessment of muscle function, a brief summary of the forms of muscle contraction, the capacities of muscle and the type of work this tissue can perform are included.

TYPES OF MUSCLE CONTRACTION

In the process of moving the parts of the body for different purposes, muscle acts in a variety of ways. It can perform movements which require the rapid development of maximal tension between the ends of the muscle. Such an action as standing on tip toes is an example of this type of response. The movement requires a concentric or shortening contraction of the gastrocnemius and soleus muscles to plantarflex the ankle joint by bringing the two attachments of the muscle closer together. Concentric contraction is generally associated with initiating movement against the resistance offered by gravity and can produce acceleration of the body segment displaced.

Muscles can also reduce the acceleration which has been imparted to a body segment either by gravity or a concentric contraction. This response of muscle is known as an eccentric or lengthening contraction. During an eccentric contraction the two attachments of the muscle are moved away from

Introduction

one another. A mother placing a baby in a cot is using the biceps brachii eccentrically to control the effects of gravity.

Where the muscle develops tension but produces no movement, the action is known as isometric contraction. An example would be provided by a child sitting on the floor who, in reaching for a toy, leans on one arm with the elbow extended. The triceps muscle on the arm providing support is working isometrically to maintain the elbow in extension and provide a stable upper extremity to aid in maintaining balance.

All the examples of behaviours used here require muscle to develop tension and are related to the strength of the muscle. The tests described in the subsequent chapters measure concentric muscle action and all are tests of strength alone. The evaluator must keep this fact in mind when interpreting the results of the manual examination.

LIMITATIONS OF MANUAL MUSCLE TESTING

As well as developing a large amount of tension, a muscle may be called upon to produce this tension at a rapid rate or to develop varying degrees of tension repeatedly over long periods. The former requires the muscle to have the ability to develop power, and the latter is an endurance ability. Neither of these abilities is measured by the manual strength testing technique. For example, the patient may score a Grade 5 on the muscle test indicating normal muscle strength, but may be lacking in endurance and therefore unable to sustain a muscle action sufficiently to perform a functional activity. Care must also be taken in interpreting the Grade 5 which indicates normal concentric muscle action but gives no measure of the capacity for eccentric action. A further problem is associated with a Grade 5 in muscles involved in activities requiring contraction against gravity and body weight (for instance, plantarflexors, hip abductors and knee extensors). The grading may not be always a useful guide to the patient's ability to stand or walk. The power required in the plantarflexors at push off in walking is much greater than that required for the standard Grade 5 test. In order to alert the tester to this problem, additional weight bearing tests have been added at the end of Chapter 4 and should be considered when appropriate to the needs of the patient.

Muscle also has capacities other than the ability to develop tension between its two attachments. These additional attributes which include excitability, extensibility, elasticity and relaxation are described in Gowitzke & Milner (1980). As manual muscle testing assesses only the capacity of the muscle to contract, it is important for the evaluator to look further than failure of contractility when seeking an explanation for the lack of an adequate response to testing.

It is beyond the scope of this text to consider direct methods of measuring these other capacities of muscle. However, some of them can be conveniently examined during the process of manual muscle testing by the simple expedient of assessing the muscle length. This manoeuvre provides information about the degree of extensibility (and conversely of relaxation) of the muscle and affords some indication of the elasticity of the passive elastic and contractile muscle elements. For this reason the position of maximum extensibility has been included with the test of strength of each muscle. The length of the muscle should always be evaluated prior to testing since abnormal length or shortening may significantly influence the outcome of the manual muscle test. More detailed descriptions of the maximal length of most muscles is to be found in Evjenth & Hamberg (1985a & b).

The problem of grading a muscle which moves a joint through more than 90° must also be considered. Such muscles as latissimus dorsi and pectoralis major are good examples. In such instances, more than one position will be required to test the whole range of the muscle for Grades 5, 4 and 3. In Part II of this book, the heading, Comments, which follows each description of the primary testing position, has been used to alert the tester when this problem arises with particular muscles.

Having stipulated the parameters of muscle action measured by muscle testing, the next step is to consider the reproducibility of this form of assessment.

STANDARDIZING THE PROCEDURE FOR MANUAL MUSCLE TESTING

In any evaluation procedure, the need to standardize the mode of assessment to ensure that the findings of the examination are reproducible is of paramount importance. Problems arise in manual muscle testing if each person grading the specific action arrives at a different value. Not only is it impossible to establish progressive rehabilitation objectives, it is demoralising for the patient to find that no two practitioners can agree. Therefore, every effort must be made to ensure that any standardized format of assessment is followed closely. Adherence to the described procedures will help to guarantee that the observation recorded by the examiner is as accurate as possible. It is the view of the present authors that if the necessary elements in the construction of any test are understood, the test user is likely to be more strongly motivated to adhere to the operating instructions which are described in the testing manual. With this idea in view, the relevant concepts related to test construction will be outlined, before presenting the format of the assessment of muscle strength.

CONCEPTS IN TEST CONSTRUCTION

In order to meet the needs of the user, any form of testing must meet a number of basic requirements. These include the three concepts of objectivity, reliability and validity, as well as the more obvious requirement that any instruments used are easy to administer, non taxing for the client and cost efficient (Barrow & McGhee 1979).

Objectivity

The first of these concepts is objectivity. In the testing situation it is essential to ensure that the results of the test are quantifiable. This imposes a number of restrictions upon the use of the test. The test user is required to follow certain rules when administering the test. These rules apply to such things as the type of equipment to be used, the physical facilities in the test situation, and the starting set-up for each of the subtests which make up the whole. To vary these rules is to reduce the likelihood that another operator using the same instrument will record results similar to those of the first tester. In terms of evaluating muscle function, the requirements for objectivity mean that the evaluator has no licence to improvise, but must follow the rules of testing faithfully and record the numerical grades which are consistent with the definitions supplied. If, as may happen, the patient's needs are such that modification of the testing requirements is necessary, all changes must be carefully documented so that in subsequent testing sessions the same modifications may be employed.

Reliability

Reliability is that characteristic of a test which allows different users to be confident that the result obtained at one evaluation will be closely related to the result obtained at a subsequent evaluation. This function is related to two facets of the concept of reliability. The first is the notion of inter-rater reliability, where different evaluators testing the same person under similar conditions will have closely comparable results. The second aspect is that of intra-rater reliability which ensures that the same tester, re-evaluating the same subject within the same time frame, will produce results which are highly correlated. The better the test construction and the more precise the instructions for administration, the more likely it is that the inter-rater and intra-rater reliability will show a positive correlation coefficient of acceptable magnitude. The correlation coefficient for these two parameters is an indication of the reliability of the test.

Validity

The third of the requirements of a test is that of validity. The concept of validity also has several components. These include face validity, predictive validity, concurrent validity and construct validity. While the methods of establishing the forms of reliability are immediately obvious from a description of the terms, the same is not true for the various forms of validity.

Face or logical validity refers to the general content of the test. Do the test items appear to be related directly to the behaviour to be evaluated? It is a fairly simple matter to glance through a manual and determine whether or not the activities described relate to evaluating the strength of specific muscles. Face validity is established by this method.

The next step is to decide what predictive value the instrument might have. If both the triceps muscles can be graded as four, then the evaluator could predict that the person with a spinal cord lesion at the level of C8 would be able to transfer from bed to wheelchair independently. Thus the measure of the muscle test's predictive validity is the degree to which the information derived from the test is an indication of the individual's functional ability. The predictive validity of the carefully administered muscle test is established with relative ease.

Concurrent validity is a concept which describes the degree of relationship between two different measures of the same behaviour recorded in the same time frame. The usual form for evaluating concurrent validity is to check any newly developed measure of a performance against a method which has long been identified as a useful measure of the activity. One way to establish the concurrent validity of manual muscle testing would be to have the patient perform the muscle action using a dynamometer and evaluate the degree of relationship between the force recorded by the dynamometer and the grade on the manual muscle test.

The last of the concepts to be considered when discussing test construction is that of construct validity. It is related to the need to establish the validity of complex acts which incorporate multiple skill parameters. Since evaluating muscle strength involves only discrete, well-defined activities, it is sufficient to establish concurrent validity only for this form of testing.

Published studies of the evaluation of objectivity, reliability and validity in manual muscle testing would appear to be lacking. Recently some authors (Baldauf et al 1984, Griffin et al 1986) have described the results of evaluations of some components of muscle testing. More extensive examination of this form of assessment would appear to be necessary. However, before addressing this problem, it is vital that the format of testing is revised so that a more uniform approach is employed. The system described here has not been vigorously examined to establish either its reliability or validity. In developing and describing it, however, every effort has been made to adhere to the principles upon which the tests of muscle strength were first constructed and to eliminate errors which have become common place over the years of clinical use. It is anticipated that this approach may help to establish the manual muscle test as an objective, reliable and valid tool in the hands of the careful assessor.

GRADING SYSTEM

A further requirement of an evaluation system is an established means of assigning a value to the muscle action elicited. Commonly used methods of grading muscle function include either a series of descriptive labels or a sequence of numerals (Medical Research Council 1943, Janda 1983, Kendall & McCreary 1983, Pact et al 1984, Daniels & Worthingham 1986).

From the point of view of data storage, analysis, retrieval and comparison, a numerical scale is much more useful. A six point numerical grading system is utilised in this text which is consistent with that of the Medical Research Council, first published in 1943. Value labels, such as 'good', 'fair' and 'poor', are excluded from the descriptions, as they are considered to be a frequent source of confusion. The evaluator's assessment is served more effectively by numerals which have a stated meaning than by commonly used value labels which are misconstrued easily.

The grades used are:

Grade 5 The muscle is able to contract through full range of joint movement against gravity, plus a resistance which is maximal for the age, sex and occupation of the subject.

Grade 4 The muscle is able to contract through full range of joint movement against gravity, plus a resistance which is less than maximal for the age, sex and occupation of the subject.

Grade 3 The muscle is able to contract through a full range of joint movement against gravity.

Grade 2 The muscle is able to contract through a full range of joint movement with gravity eliminated.

Grade 1 A muscle contraction is detectable on palpation. Movement of the joint is minimal or absent.

Grade 0 No muscle contraction is detectable on palpation.

The use of 'plus' and 'minus' within this grading system is a form of supplementation which many clinicians employ. Careful thought has been given to this practice by the present authors. Despite common usage, negatives have been excluded from the grading system used here as, for example, there is no logic in describing a muscle as having a Grade 5 minus strength. It can either produce a full range of movement against a maximal resistance or it cannot. If the muscle strength does not meet the criteria for a Grade 5, it should be graded as a 4. On the other hand, there is some basis for the incorporation of a positive sign. The plus can be used to assist in differentiating between the strength of a Grade 1 and Grade 2. A plus could be added to Grade 1 when the strength of the muscle produces some joint movement but it is not through full range. A Grade 2 plus can be given when the muscle can move the joint through full range with some resistance, but not through a full range of movement against gravity. The use of these grades covers the majority of possible situations encountered in muscle testing.

In presenting the evaluation of the muscles of mastication, facial expression and essential functions, the six point grading scale is questioned. The value of introducing a separate scale is also thought to be minimal. A compromise has been struck which employs the six point scale but omits those grades which would appear to have little meaning. Therefore, a muscle which functions normally through its full range of movement is graded as 5. Where the muscle naturally acts against a constant resistance which is uninfluenced by gravity, manual resistance is not applied. A constant testing position is employed which mimics closely a usual functional position. Where the muscle activity is less than normal a Grade 2 is assigned. As is the case with other muscles, when there is no muscle activity a Grade 0 is assigned.

Where the grading system does not adequately describe the test result, alternative descriptions can be used.

ACCESSORY DESCRIPTORS

Frequently the inability of a muscle to contract through its full range is a result of factors other than or additional to muscle weakness. In order to preserve the reliability of the manual muscle testing procedure it is better that these instances are dealt with using an accessory descriptor. This prevents the grading system being used incorrectly and makes the evaluator attend more closely to the problem which interferes with the muscle action.

Accessory descriptors are listed below and can be entered on the test form beside the muscle in question in place of a grade.

P: the presence of pain inhibiting the complete muscle action.
T: the existence of abnormal muscle tone inhibiting the complete muscle action.
ROM: either limited or excessive range of joint motion inhibiting the complete muscle action.
C: muscle or soft tissue contracture inhibiting the complete muscle action.
PS: the loss of proximal stability inhibiting the complete muscle action.

F: failure to produce the full muscle action as a result of the patient's lack of understanding or inability to co-operate.

OM: the test for that muscle was omitted.

Where any of these descriptors are utilised, a full explanation of their meaning in the context of the subject's performance must be included in the space provided for comments on the muscle testing form. For example, if the subject is reluctant to initiate shoulder abduction against gravity because of the occurrence of pain in the 0–30° range but can achieve at the Grade 3 level once beyond that point, it is noted that the total action cannot be graded as a 3. In this instance both the supraspinatus and middle fibres of deltoid should be assigned a P and the nature of the inhibition detailed. It is useful also to indicate the range of joint movement which is affected. Thus it might be recorded as a Grade 2P from 0–30°, and a Grade 3P from 30–90°. The use of the accessory descriptors prevents the abuse of the grading system and requires the evaluator to record accurately any deviation in performance noted during the testing session.

SEQUENCE OF PRESENTATION

Every effort has been made in this text to produce a consistent format in the description of the testing procedures. The muscles which contribute to postural stability and movement have been identified and appropriate tests described. A list of those involved in each action is provided at the beginning of each section, together with a photograph which identifies the muscles superficially. Each muscle is then described in terms of its gross attachments, nerve supply, surface marking for palpation, actions, position of maximum extensibility, and testing procedures. Where there is a marked difference between the positioning of the subject for testing against gravity and with gravity eliminated, both situations are depicted. Where the position varies little, or is consistent over a number of muscles which move the same segment, the gravity eliminated position is either omitted or shown only once for a sequence of muscles.

Muscles which share common innervation and actions and cannot be differentiated adequately in testing are identified separately, but grouped together in those elements which they share.

ADDITIONAL ELEMENTS

Two concepts are presented which are not usually part of a muscle testing text. Firstly, the idea of the dual actions of a muscle which performs work both by moving its caudal (distal) attachment closer to the rostral (proximal) attachment and by pulling in the reverse direction when the caudal segment is fixed by the body's weight. Thus the actions performed by each muscle with first the rostral and then the caudal end fixed are described. The second concept is related to recognising the extent to which a muscle can be fully elongated, a capacity which is indicative of the extensibility of both the contractile and non-contractile elements of the muscle. The position of maximum extensibility has been included with the test of strength of each muscle. The length of the muscle should always be evaluated prior to testing, since abnormal length or shortening may significantly influence the outcome of the manual muscle test. More detailed descriptions of the maximum length of most muscles is to be found in Evjenth & Hamburg (1985 a & b). Inclusion of these additional elements in the evaluation of muscle function is considered to aid the evaluator in identifying factors likely to contribute to any diminished muscle function observed.

For some muscles, however, muscle action in terms of caudal or rostral fixation, extensibility, and surface markings for palpation, have no meaning. This is the case when it is not possible to stretch the muscle manually, superficial palpation is not possible, the muscle's function is not sufficiently understood to consider a reversal of that function, or there is no reversed action (as is the case for many of the facial muscles). Components of the sequence of presentation are redundant therefore, in some instances. Where this is the case, irrelevant headings have been omitted.

In addition, by including a chapter on the muscles involved in essential functions, it is hoped to alert the user to the need to assess comprehensively those muscles which provide a different behavioural dimension. It is just as important to assess the action of the pelvic diaphragm as it is to assess more easily visualised muscles.

Having identified all the traps which tend to reduce the usefulness of manual muscle testing as a measure of performance, it is appropriate to now discuss the clinical application of the testing process.

The muscle test is an essential component in the total assessment of patients with a wide variety of movement abnormalities. It may be employed as an aid to diagnosis of the pathology of the emerging condition or it may form part of an ongoing recovery assessment after a period of treatment. In preparing to test muscle function it is important to consider the testing environment. An optimal environment increases the likelihood that the subject will be able to demonstrate a maximal performance which, after all, is the objective of the testing procedure.

OPTIMAL PREPARATION FOR TESTING

Because the patient should be suitably undressed during the testing, it is important that the temperature of the room is comfortable. An overheated environment increases the rate of muscle fatigue, while a room that is too cold may cause the patient to shiver and develop increased tension. Both situations will interfere with optimal performance. Again, because the patient is undressed, the testing area should afford privacy. Additionally, a quiet environment will allow full concentration on the task in hand and minimise the adverse effects caused by distractions and interruptions.

The space available should be adequate for the task so that the test can proceed smoothly without the need to rearrange furniture. A chair and, where possible, an adjustable couch; pillows, linen, recording chart, and any other items such as a spoon, plate, ice cube or food required for the evaluation of swallowing, should be prepared in advance. The tester should set aside a block of uninterrupted time, sufficient for the task, so that the session is appropriately paced to suit the patient's needs. A hurried test or a distracted tester both diminish the patient's ability to demonstrate maximum performance.

Two repetitions of each action are sufficient for adequate testing. The best performance should be systematically noted after each test to ensure an accurate recording is made. No test should be repeated more than three times. If there are doubts about the performance of a muscle after two repetitions, another muscle should be evaluated while the doubtful muscle is rested. A third repetition can then be attempted when the suspect muscle has had time to renew its anaerobic energy resources.

It is important that the patient understands the purpose of testing and comprehends the tester's instructions. Therefore, the tester must take the trouble to explain each task sufficiently to ensure maximum co-operation, and offer a demonstration for the patient to copy, if required.

In order to make the most efficient use of time, a logical order of test item presentation should be sought. Following a consistent format has the added advantage of aiding memory and enabling the tester to increase efficiency with repeated use of the test procedure. The advantage from the patient's viewpoint is the saving in energy resulting from limiting the number of position changes required in completing the testing sequence. Based on a preliminary estimate of the patient's ability to perform movements against gravity, it is common for the therapist to commence by testing in the gravity resisted positions. The suggested sequence for examining Grades 5, 4 and 3 in such patients is outlined in Table 2.1 (p. 21). For patients who are unable to achieve at the Grade 3 level or above, the therapist will need to choose an alternative sequence for testing (see Table 2.2, p. 22).

The sequences outlined in these two tables begin by examining all muscles which can be tested in the supine lying position and progress through half lying, side lying, the prone position, sitting, and standing. By following a logical sequence and attending to detail in this way, the process is stream-lined, the patient is less likely to be fatigued, and the outcome of the testing session will be enhanced.

LEGITIMATE MODIFICATIONS

Despite the best efforts of the conscientious tester, the needs of many patients are such that the testing procedures may require modification on occasions. Such modifications do not invalidate the test provided they are appropriately recorded. By indicating the changes made to meet the patient's needs, the tester ensures that the conditions which prevailed at the time of testing can be reproduced, if required, in subsequent testing sessions. Common instances in which modifications may be necessary, include the testing of patients who have diminished or absent peripheral sensation, obese individuals, the confused and disorientated patient, the perceptually impaired patient, children, the elderly, and intellectually handicapped people.

Sensory loss

Where sensory loss is a problem, it is essential that the patient can utilize vision to help overcome the absent or inadequate proprioceptive or tactile input. Thus the position employed to test some muscles may need altering to enable the patient to see the required movement. In changing the testing position to accommodate the patient, the relationship to gravity will be altered. This must be taken into account when assigning a grade to the muscle following testing in the modified position. Failure to account for the change in position in grading may leave subsequent evaluators with an invalid perception of the previous abilities of the patient.

Obesity

The obese individual presents special problems in testing. The cover of adipose tissue will complicate palpation of the muscle to be tested and make identification of surface markings difficult. There is no simple answer to the difficulties encountered. Obviously, bony landmarks will prove more reliable than soft tissue ones, but in order to penetrate the superficial layers of tissue to identify these landmarks, the palpation must be of sufficient pressure. This degree of pressure may be painful or uncomfortable and can produce bruising, therefore care in palpation is essential. In addition, the range of movement demonstrated by the patient may be diminished because the additional tissue layers increase the circumference of the limbs and cause adjacent segments to come into contact before the full range of joint movement has been attained. Similarly, the amount of tension which must be developed in muscles to move segments which carry additional adipose tissue may be greater than would normally be required to move the same segment in a person of average weight. The fact that the range of movement through which the patient can move is less than the average but is a full range for this individual, should be noted. The added weight of the limb should not, however, be used to replace external resistance. The testing procedures are based upon the application of resistance for the establishment of Grades 4 and 5, therefore manual loading must be employed in order to retain consistency in testing.

Confusion and disorientation

Where the patient is confused and disorientated testing sessions should be short. The effort of attending to instruction will produce fatigue rapidly in this type of patient. The chances of achieving co-operation will be greater before fatigue develops and testing may be more successful if carried out over several, shorter periods. The confused state of the patient should be noted and where doubt exists as to the reliability of a response, this should also be recorded.

Perceptual dysfunction

The patient who has suffered central nervous system damage producing perceptual dysfunction, presents a major challenge to the tester. The challenge is to differentiate between loss of muscle power, and a variety of perceptual problems such as receptive aphasia, apraxia, neglect or somatognosia likely to inhibit motor function. Observation of automatic movements such as postural reactions to perturbation of the patient's centre of gravity may provide clues to the nature of the problem and should be utilized in seeking a solution to the dilemma such patients present. Once it is established that perceptual dysfunction is the principal problem, then it is inappropriate to continue the manual muscle test as the results are unlikely to be reliable or to be a valid measure of muscle function in that person.

The elderly patient

For the elderly patient, fatigue will frequently be the principal factor limiting performance. Other age-related problems, such as diminished auditory and visual acuity, may also interfere with the older person's ability to co-operate. Auditory loss makes it difficult for the tester's instructions to be heard and the visual loss reduces the capacity of the patient to benefit from a demonstration. Older patients are more susceptible to fatigue because of limited physical endurance and testing should be confined to shorter periods and perhaps carried out over several sessions.

In dealing with the diminishing sensory acuity of the elderly person, it is important to ensure that the patient can observe the therapist's face readily. This allows the patient to use clues such as facial expressions to augment auditory cues, and provides a focus for directing attention to the required muscle. Tactile cues may also be employed to help patients centre their efforts on the component being examined.

Young children and the intellectually handicapped

Manual muscle testing in young children and individuals who are intellectually handicapped presents difficulties which are similar for both groups. Neither group is able to co-operate in the testing situation, and both may prove reluctant to participate. Testing in the neonatal period is often required for infants who suffer nerve lesions as a result of the delivery process or who manifest congenital abnormalities at birth. Erb's palsy or spina bifida cystica are examples of neonatal pathologies requiring evaluation of muscle function in the first weeks of life. Fortunately, infants respond readily to tactile stimuli and light stroking over the belly of a muscle will often produce the desired contraction.

It is possible in neonates to establish that a muscle has the capacity to contract, but grading through the whole scale is not feasible. Remembering that even infants who have no abnormalities will be unable to produce a contraction against graded resistance because of the immaturity of the neuromuscular complex, it is reasonable to grade the muscle as a 5 if a full range of movement against gravity can be elicited. Where less than a full range can be observed a Grade 2 would be a more useful grade, and a 0 grade is appropriate when no response can be elicited.

For toddlers in the 12 month–4 year age range, testing requires a good deal of ingenuity. Results obtained in this age group are likely to be the least reliable of all (McDonald et al 1986). Many young children vigorously resist handling by strangers and are reluctant to undress. Both of these factors mean that the child is upset easily and the observation and palpation are made quite difficult. Results obtained in a crying child are unlikely to be optimal and the situation should be avoided. It may be possible to have the mother lead the child through a series of activities which will provide a guide to muscle strength but the reliability of this type of evaluation is dependent to a large extent upon the experience of the therapist in dealing with children. Table 2.3 (p. 23) lists a sequence of activities which the child may be encouraged to perform and which can provide basic information about a wide range of muscles. The use of the sequence may prove to be a useful alternative when handling the child proves to be impossible.

From the age of 5 onwards children can be tested formally using the same procedures as for adults. However, because many of the muscle actions tested in the adult require a high degree of motor skill to produce these actions in isolation, children may have difficulty with a large number of the test items. Particularly difficult for young children are tests which require isolated movements of the fingers and toes. These are frequently beyond the competence of 5-year-olds and inability to perform on these items does not indicate specific muscle weakness. Young children, like older persons, will have difficulty in concentrating for long periods of time. Testing should be undertaken, therefore, in repeated short sessions. Alternative approaches to evaluating muscle strength in children have been suggested by various authors. For further information, the reader is referred to Alexander & Molnar (1973), Molnar & Alexander (1974), Baldauf et al (1984), Pact et al (1984), and Lefkok (1986).

Many intellectually handicapped persons are able to co-operate sufficiently to permit standard test procedures to be employed. The amount of co-operation is dependent on the therapist's experience with such individuals and the degree of the intellectual disability. Depending on the age and size of the patient, activities such as those outlined for toddlers in Table 2.3 may be helpful in establishing the presence or absence of muscle function where it is not possible to undertake an accurate evaluation.

SPECIAL PROBLEMS

Evaluating muscle strength in the presence of abnormal muscle tone is the subject of some controversy. Many therapists believe that muscle weakness is not a feature of upper motor neuron lesions and do not include muscle strength testing in the evaluation of patients with these disorders. However, the amount of tension produced by a muscle depends not only on the size and number of motor units activated, but also on the frequency at which they are stimulated and the synchrony of recruitment. The cerebral cortex is largely responsible for the organization of these neural recruitment processes. Also the type of muscle fibre, its cross-sectional area, and the condition of both the neuromuscular junction and the muscle itself, influence the amount of tension produced by the muscle (Sanders & Sanders 1985). It is probable then that secondary loss of muscle strength is one of the features of the motor dysfunction of patients suffering from central nervous system disorders in which muscle tone is altered. The manual muscle test is a valuable adjunct to the evaluation of the abilities of such patients. However, the problems of fluctuating tone and limited range of movement due to increased tone will further complicate the assessment and should be carefully noted.

Frequently, the individual recovering from a stroke or head injury presents with both increased tone and weakness in the same muscle. Where abnormal tone causes difficulty in assigning a grade to a tested muscle, the appropriate Accessory Descriptor should be employed instead of a grade and an outline of the problem included in the space provided on the muscle grading chart (see Appendices 1 and 2).

The acutely ill patient presents another series of problems to the therapist who is asked to evaluate muscle strength. The person who has been admitted to hospital with Guillain-Barré syndrome will often require careful assessment to establish the stage of progression of the paralysis. Apart from the fact that both central and peripheral fatigue may be a feature at this stage of the disease, the patient in an effort to co-operate may unwittingly substitute alternate muscle actions for that of the paralysed muscle, producing trick movements. Trick movements occur most commonly because of failure to position accurately the component to be moved, or through lack of stabilization of the joints on either side of the movement. Adequate stabilization to isolate movement correctly is of particular importance when testing the muscles which move the wrist, fingers and thumb, and the ankle and toes. The weaker the muscle to be evaluated the more important proximal stabilization becomes.

The patient who demonstrates nerve root signs frequently complains of pain and will also require special care. A movement which is painful may reduce the grading assigned to a muscle despite the fact that the tester suspects the strength to be normal. It is often helpful, where pain is a feature of the patient's condition, to ensure that appropriate analgesic cover is established prior to testing in order to facilitate maximum co-operation in the testing procedure.

AIDS TO TESTING AND DIAGNOSIS

The most common reason for utilizing manual muscle testing is associated with isolated lesions of the peripheral nerves. Testing in these cases can be

Clinical guidelines

limited to the muscles supplied by the nerve trunk involved since the damage is selective.

To assist the clinician in discrete testing such as this, a tabular presentation of the muscles supplied by the three cords of the brachial plexus and the major nerve trunks of the lumbosacral plexus has been devised. Figure 2.1 (p. 18) presents the first of the three components of the brachial plexus: the posterior cord. The medial and lateral cords of this plexus are combined in Figure 2.2 (p. 19). The nerves which arise from the separate cords and the segmental contribution the spinal cord makes to each are shown in **bold** type. The muscles innervated by these nerves are listed in the order in which branches supplying them are given off.

A therapist called upon to investigate loss of motor function sustained when a lesion occurs in one of the nerves arising from the brachial plexus can begin to isolate the problem by observation. For example, loss of the axillary nerve produces a characteristic inability to abduct or flex the glenohumeral joint due to the lack of innervation of the deltoid muscle. Radial nerve palsy may be accompanied by the loss of wrist and finger extension and forearm supination. This produces the easily detectable dropped wrist phenomenon. Where damage to the radial nerve occurs in the axilla, active elbow extension will be lost in addition to the distal musculature. The patterns of muscle atrophy which accompany ulnar and median nerve damage are also very identifiable. The ape-like hand produced by the loss of opposition which occurs in median nerve lesions, and the clawed appearance of the hand which results from interosseus loss in ulnar nerve palsy, should be readily observable.

Where such signs are not present, the history of the onset of muscle weakness can provide valuable clues to the likely cause of the dysfunction. The location of a fracture may give an anatomical marker for the nerve supply likely to have been damaged. To pursue the example of the radial nerve involvement, a common injury likely to damage this nerve is a fracture of the mid-shaft of the humerus. The radial nerve is interrupted as it spirals around the shaft of the humerus.

Alternately, a gradual onset of weakness may suggest compression. Likely sites for this to occur include the intervertebral foramina. Other signs of nerve root involvement accompanying the history of gradual onset may implicate the segmental input to the cords of the brachial plexus. Such damage would produce a diffuse loss of function and would require accurate testing of more than one peripheral nerve to map the extent of the paralysis.

By first utilizing information sources such as the history of onset, and the type and location of injury, the nerve supply likely to be involved can be identified. The tester should select the appropriate nerve from either Figures 2.1 or 2.2 and evaluate the performance of the muscles listed. The resultant grading, recorded in the muscle grading chart, will provide a description of the extent of muscle paralysis or weakness developing as a consequence of the neural pathology.

Figures 2.3 and 2.4 (pp. 19–20) provide a similar guide for the examination of muscles innervated by the lumbosacral plexus. Traumatic injuries to the nerve supply of the lower extremities are less common. However, those which do occur tend to be more diffuse in effect because of the nature of the causative agents. Common causative agents are motor vehicle accidents and gunshot wounds. The resulting loss of muscle function can be relatively unique for many patients; thus it is essential to evaluate strength carefully in terms of the segmental roots and cords as well as the nerves supplying individual muscles, in order to localize correctly the damaged neural components. Since the injuries are less common than those occurring in the upper extremity and are also more unique, the pattern of paralysis is not as easily recognised. However, the tester, using Figures 2.3 and 2.4 as a guide, should be able to develop a logical description of the neural damage.

After peripheral nerve lesions, the most common application of manual muscle testing is in patients who have a spinal cord injury. For such patients, the establishment of the level of the lesions is a valuable pointer in choosing rehabilitation goals. To the clinician experienced in evaluating muscle function, the task is routine. To the beginning practitioner, deciding where to start can be a daunting task.

In order to understand the consequences of cord damage, the tester must have a clear picture of the description of the level of the lesion. Confusion arises in understanding the level of the lesion because the spinal cord is shorter than the vertebral column, in the adult. Therefore the nerve roots exiting through the vertebral foramina in the thoracic, lumbar and sacral areas which originate in correspondingly numbered segments of the cord, are formed from at least two vertebral levels above the point of exit. Bromley (1985) describes the most common classification of the lesion levels. To describe the motor level of the lesion, the most distal uninvolved level of the cord is named. Thus, a lesion complete below T11 indicates that all muscles supplied by T11 and above will have normal function. The level of the skeletal lesion, that is, the level at which the fracture or dislocation of the vertebrae has occurred, will be above this at T9–10. Tables 2.4, 2.5 and 2.6 use this definition of the level of lesion and should be used by the practitioner both to decide where to begin in the examination of the spinal cord injured patient and to identify the motor levels of the spinal cord which are significant to the establishment of rehabilitation goals.

The innervations of muscle described in *Gray's Anatomy* (Williams & Warwick 1980) have been used in developing Tables 2.4, 2.5 and 2.6 (see pp. 24–27). The rationale for these Tables is based upon the fact that muscles which are innervated by more than one segment of the spinal cord will display minimal or partial control when the damage occurs at the rostral level of the nerve supply to that muscle, and complete control when the level of the cord lesion lies below the principle innervation of the identified muscle. Muscles likely to demonstrate complete control are listed on the left-hand side of the Tables. Those in which all function is likely to be minimal are listed on the right and the central column lists muscles likely to have retained some function. Table 2.4 deals with the muscles which move the neck and upper limb, Table 2.5 with those that move the trunk, and Table 2.6 with the muscles responsible for movement of the lower limb.

MUSCLE GRADING CHARTS

Two charts for recording the results of the manual muscle test are included as Appendices 1 and 2. The first includes the muscles of the trunk and extremities, the second covers those of the face and essential functions. For ease of use the numerical grades and the accessory descriptors employed are listed and defined on both, and space for patient details has been provided on the first page. Additionally, sections for relevant comments about the performance of the muscles tested have been included at the end of each chart. It is on this page that any necessary modifications employed or particular problems encountered should be carefully outlined.

The muscle grading charts are arranged so that the muscles, the names of the nerves supplying them, and the level of spinal cord innervation are shown in the central column. On either side are a series of smaller columns in which the grading assigned to each muscle can be recorded. The charts are designed to be used on multiple occasions so that a picture of recovery, degeneration, or static performance can be seen easily with repeated testing. The date of each test is entered above the recording column as indicated and the same chart used at subsequent testing occasions so that a full account of the patient's progress is maintained in the one document. To further simplify the recording process, the muscles on the left-hand side of the body are recorded on the left-hand side of the chart and those muscles of the right-hand side of the body appear on the opposite side of the chart.

The order of presentation of muscles in the charts is not the same as that described in Tables 2.1 and 2.2. Unfortunately, to list the muscles on the muscle grading charts in the suggested sequence of testing would limit the value of the charts as a diagnostic tool. Therefore, the muscles of the trunk and limbs have been grouped according to their level of innervation so that the location of the lesion will be apparent from the gradings assigned to muscles which receive their nerve supply from contiguous levels of the spinal cord. The muscles associated with essential functions have been grouped according to those functions.

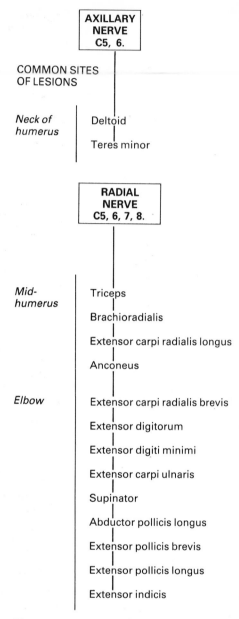

Fig. 2.1 Nerves from the posterior cord of the branchial plexus

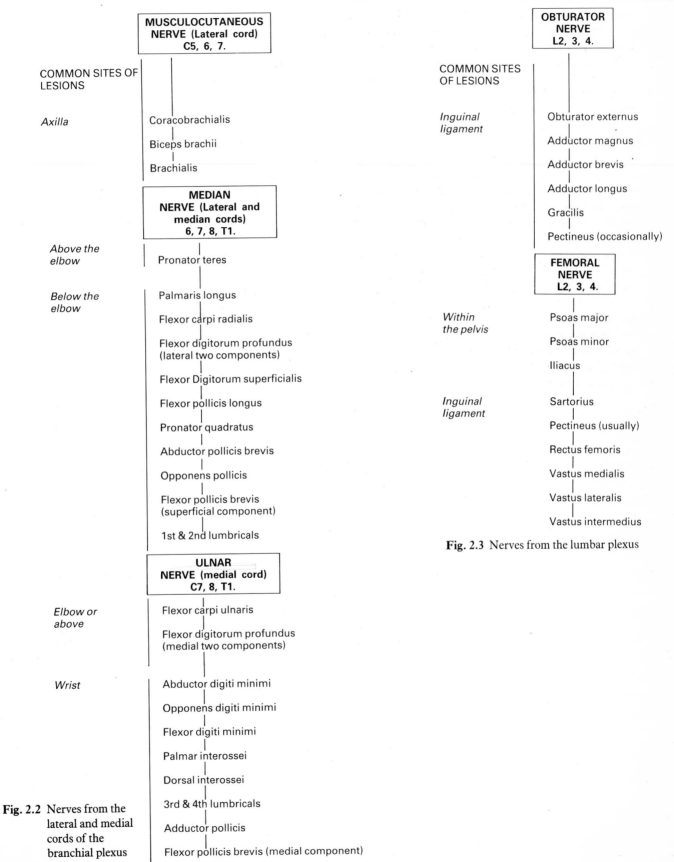

COMMON SITES OF LESIONS

	MUSCULOCUTANEOUS NERVE (Lateral cord) C5, 6, 7.
Axilla	Coracobrachialis
	Biceps brachii
	Brachialis

MEDIAN NERVE (Lateral and median cords) 6, 7, 8, T1.

Above the elbow — Pronator teres

Below the elbow
- Palmaris longus
- Flexor carpi radialis
- Flexor digitorum profundus (lateral two components)
- Flexor Digitorum superficialis
- Flexor pollicis longus
- Pronator quadratus
- Abductor pollicis brevis
- Opponens pollicis
- Flexor pollicis brevis (superficial component)
- 1st & 2nd lumbricals

ULNAR NERVE (medial cord) C7, 8, T1.

Elbow or above
- Flexor carpi ulnaris
- Flexor digitorum profundus (medial two components)

Wrist
- Abductor digiti minimi
- Opponens digiti minimi
- Flexor digiti minimi
- Palmar interossei
- Dorsal interossei
- 3rd & 4th lumbricals
- Adductor pollicis
- Flexor pollicis brevis (medial component)

Fig. 2.2 Nerves from the lateral and medial cords of the branchial plexus

COMMON SITES OF LESIONS

	OBTURATOR NERVE L2, 3, 4.
Inguinal ligament	Obturator externus
	Adductor magnus
	Adductor brevis
	Adductor longus
	Gracilis
	Pectineus (occasionally)

FEMORAL NERVE L2, 3, 4.

Within the pelvis
- Psoas major
- Psoas minor
- Iliacus

Inguinal ligament
- Sartorius
- Pectineus (usually)
- Rectus femoris
- Vastus medialis
- Vastus lateralis
- Vastus intermedius

Fig. 2.3 Nerves from the lumbar plexus

Clinical guidelines

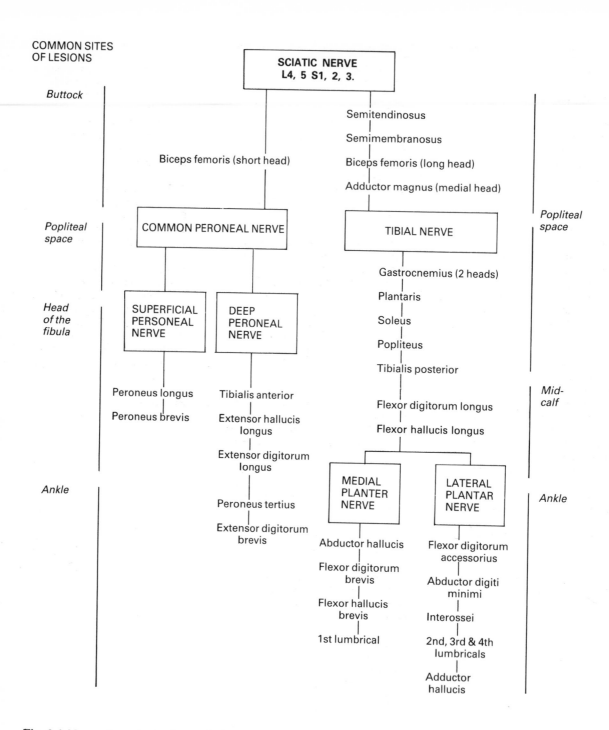

SCIATIC NERVE
L4, 5 S1, 2, 3.

Buttock

Biceps femoris (short head)

Semitendinosus

Semimembranosus

Biceps femoris (long head)

Adductor magnus (medial head)

Popliteal space

COMMON PERONEAL NERVE

TIBIAL NERVE

Popliteal space

Gastrocnemius (2 heads)

Plantaris

Head of the fibula

SUPERFICIAL PERSONEAL NERVE

DEEP PERONEAL NERVE

Soleus

Popliteus

Tibialis posterior

Peroneus longus

Peroneus brevis

Tibialis anterior

Extensor hallucis longus

Flexor digitorum longus

Flexor hallucis longus

Mid-calf

Extensor digitorum longus

Peroneus tertius

Ankle

MEDIAL PLANTER NERVE

LATERAL PLANTAR NERVE

Ankle

Extensor digitorum brevis

Abductor hallucis

Flexor digitorum brevis

Flexor hallucis brevis

1st lumbrical

Flexor digitorum accessorius

Abductor digiti minimi

Interossei

2nd, 3rd & 4th lumbricals

Adductor hallucis

Fig. 2.4 Nerves from the lumbosacral plexus

Table 2.1 Positions for testing Grades 5, 4 and 3 muscle strength. (Alternative testing position shown in brackets)

Supine lying
Lumbricals
Plantar interossei
Adductor hallucis
Dorsal interossei
Abductor digiti minimi
Abductor hallucis
Flexor digitorum longus
Flexor digitorum brevis
Flexor digitorum accessorius
Flexor hallucis longus
Flexor hallucis brevis
Extensor hallucis longus
Extensor hallucis brevis
Extensor digitorum longus
Extensor digitorum brevis
Peroneus tertius
Psoas, middle and outer range
Iliacus, middle and outer range
Rectus abdominis ⎫
Obliquus externus |
 abdominis } Abdominals
Obliquus internus |
 abdominis ⎭
Neck flexors
Sternocleidomastoid
Pectoralis major, outer and middle
 ranges
Pectoralis minor
Serratus anterior
Latissimus dorsi, outer and middle
 ranges
Teres major, outer to middle range
Subscapularis, outer to middle range
Teres minor, outer to middle range
Infraspinatus, outer to middle range
Anconeus
Triceps

**Side lying quarter turned towards
 supine**
Tensor fascia latae

Side lying
Gluteus medius
Adductor magnus, longus, brevis
Pectineus
Gracilis
Neck lateral side flexors
Pelvic floor

Prone lying
Gastrocnemius
Plantaris
Soleus
Biceps femoris ⎫
Semitendinous } Hamstring group
Semimembraneous ⎭
Gluteus maximus (inner range)
Back extensors
Neck extensors
Rhomboids — major and minor
Trapezius — middle and lower fibres
Latissimus dorsi, middle to inner range
Teres major, middle to inner range
Subscapularis, middle to inner range
Infraspinatus, middle to inner range
Teres minor, middle to inner range

Sitting
Peroneus longus (also side lying)
Peroneus brevis (also side lying)
Tibialis posterior (also prone lying)
Tibialis anterior (also supine lying)
Popliteus
Quadriceps
Quadratus femoris, middle to inner range
Piriformis, middle to inner range
Gemelli — Inferior, middle to inner range
 — Superior, middle to inner
 range
Obturator internus, middle to inner range
Obturator externus, middle to inner range
Gluteus minimus, middle to inner range
Iliposoas, inner range
Levator scapula
Trapezius — upper fibres
Coracobrachialis
Deltoid — anterior fibres (also supine)
Deltoid — posterior fibres (also prone)
Deltoid — middle (also side lying)
Supraspinatus (also side lying)
Biceps brachii
Brachioradialis
Brachialis
Supinator
Pronator teres
Pronator quadratus
Palmaris longus
Flexor carpi ulnaris
Flexor carpi radialis

Extensor carpi radialis longus
Extensor carpi radialis brevis
Extensor carpi ulnaris
Lumbricals
Flexor digiti minimi
Flexor digitorum superficialis
Flexor digitorum profundus
Abductor digiti minimi
Dorsal interossei
Palmar interossei
Extensor digitorum
Extensor indicis
Extensor digiti minimi
Opponens digiti minimi
Flexor pollicis brevis
Flexor pollicis longus
Extensor pollicis brevis
Extensor pollicis longus
Abductor pollicis brevis
Abductor pollicis longus
Adductor pollicis
Opponens pollicis
Facial muscles
Swallowing

Standing
Quadratus lumborum

Clinical guidelines

Table 2.2 Positions for testing Grades 2, 1 and 0 muscle strength. (Alternative testing position shown in brackets)

Supine lying
Lumbricals (also half lying)
Plantar interossei (also half lying)
Adductor hallucis (also half lying)
Dorsal interossei (also half lying)
Abductor digiti minimi (also half lying)
Abductor hallucis (also half lying)
Flexor digitorum longus (also half lying)
Flexor digitorum brevis (also half lying)
Flexor digitorum accessorius
Flexor hallucis longus (also half lying)
Flexor hallucis brevis (also half lying)
Extensor digitorum longus (also half lying)
Extensor digitorum brevis (also half lying)
Extensor hallucis longus (also half lying)
Extensor hallucis brevis (also half lying)
Peroneus longus (also half lying)
Peroneus brevis (also half lying)
Tibialis posterior (also half lying)
Tibialis anterior (also side lying)
Peroneus tertius (also half lying)
Quadratus femoris
Piriformis
Gemelli — inferior
        — superior
Obturator internus
Obturator externus
Gluteus minimus
Gluteus medius
Adductor magnus
Adductor brevis
Adductor longus
Pectineus
Gracilis
Levator scapula
Deltoid — middle fibres
Supraspinatus

Side lying
Extensor digitorum longus (also supine lying)
Extensor digitorum brevis (also supine lying)
Extensor hallucis longus
Extensor hallucis brevis
Tibialis posterior
Plantaris
Gastrocnemius
Soleus
Rectus femoris ⎫
Vastus lateralis ⎪ Quadriceps femoris
Vastus medialis ⎬
Vastus intermedius ⎭
Biceps femoris ⎫
Semitendinosus ⎬ Hamstring group
Semimembranosus ⎭
Psoas
Iliacus

Gluteus maximus
Rectus abdominis ⎫
Obliquus externus ⎪
  abdominis ⎬ Abdominals
Obliquus internus ⎪
  abdominis ⎭
Back extensors
Neck flexors
Neck extensors
Latissimus dorsi
Coracobrachialis
Deltoid — anterior fibres
Deltoid — posterior fibres
Teres minor
Infraspinatus
Teres major
Subscapularis

Side lying quarter turned towards prone
Tensor fascia latae

Prone lying
Quadratus lumborum
Trapezius-upper fibres
Trapezius-lower fibres

Sitting
Popliteus
Obliquus externus ⎫
  abdominis ⎪
  ⎬ Abdominals
Obliquus internus ⎪
  abdominis ⎭
Sternocleidomastoid
Rhomboids-major and minor
Trapezius-middle fibres
Pectoralis minor
Serratus anterior
Teres major
Subscapularis
Teres minor
Infraspinatus
Pectoralis major — clavicular head
Biceps brachii (also side lying)
Brachialis (also side lying)
Brachioradialis (also side lying)
Triceps (also side lying)
Supinator (also supine lying)
Pronator (also supine lying)
Palmaris longus (also supine lying)
Flexor carpi ulnaris (also supine lying)
Flexor carpi radialis (also supine lying)
Extensor carpi radialis longus (also supine lying)
Extensor carpi radialis brevis (also supine lying)
Extensor carpi ulnaris (also supine lying)
Lumbricals (also supine lying)
Flexor digiti minimi (also supine lying)
Flexor digitorum superficialis (also supine lying)

Flexor digitorum profundus (also supine lying)
Abductor digiti minimi (also supine lying)
Dorsal interossei (also supine lying)
Palmar interossei (also supine lying)
Extensor digitorum (also supine lying)
Extensor indicis (also supine lying)
Extensor digiti minimi (also supine lying)
Opponens digiti minimi (also supine lying)
Flexor pollicis brevis (also supine lying)
Flexor pollicis longus (also supine lying)
Extensor pollicis brevis (also supine lying)
Extensor pollicis longus (also supine lying)
Abductor pollicis brevis (also supine lying)
Abductor pollicis longus (also supine lying)
Adductor pollicis (also supine lying)
Opponens pollicis (also supine lying)
Facial muscles (also half lying)
Swallowing

Table 2.3 Toddlers' tests

Activity	Muscles assessed
Ventral suspension	Neck, trunk and hip extensors
Aeroplanes in ventral suspension	Trapezius (middle and lower fibres), rhomboids, teres minor, infraspinatus
Rolling	Pectoralis major, deltoid (anterior fibres), neck side flexors, hip flexors, abdominals
Squatting	Gluteus maximus, quadriceps, back extensors
Prone lying to sitting	Trunk and neck extensors and side flexors, triceps, serratus anterior
Supine lying to sitting	Neck flexors, abdominals, trunk side flexors
Creeping	Hip adductors and flexors, hamstrings, triceps, serratus anterior, neck extensors
Downward parachute	Ankle dorsiflexors, quadriceps, hip extensors, hip abductors
Walking	Hip abductors, quadriceps, hamstrings, plantarflexors and dorsiflexors, evertors
Toe walking	Ankle plantarflexors
Heel walking	Ankle dorsiflexors and evertors
Grasping a toy	Finger flexors, palmar interossei, wrist extensors
Waving goodbye	Wrist extensors and finger flexors and extensors
Picking up a sultana (pincer grasp)	Opponens pollicis, wrist extensors
Pointing	Extensor indicis
Protective extension forwards	Dorsal interossei, wrist and elbow extensors, shoulder flexors, serratus anterior
Protective extension to the side	Finger, wrist and elbow extensors, shoulder abductors

Clinical guidelines

Table 2.4 Motor function in cervical cord lesions

LEVEL	1. COMPLETE CONTROL	2. PARTIAL CONTROL	3. MINIMAL CONTROL
C3	Infrahyoids Rectus capitis, anterior & lateral Longus capitis Sternocleidomastoid	Neck extensors Trapezius Levator scapulae	Diaphragm Scaleni Longus colli
C4	**All of column 1 plus** Trapezius	Neck extensors Longus colli Diaphragm Levator scapulae	Scaleni
C5	**All of column 1 plus** Diaphragm Rhomboids Supraspinatus Levator scapulae Teres minor	Neck extensors Scaleni Longus colli Infraspinatus Subscapularis Teres major Deltoid Biceps Brachialis Brachioradialis	Pectoralis major
C6	**All of column 1 plus** Subclavius Infraspinatus Subscapularis Teres major Deltoid Biceps Brachialis Brachioradialis Supinator	Neck extensors Longus colli Scaleni Serratus anterior Coracobrachialis Extensor carpi radialis, longus & brevis Pronator teres Palmaris longus Abductor pollicis longus Extensor pollicis brevis Flexor carpi radialis Abductor pollicis brevis Opponens pollicis Flexor pollicis brevis 1st & 2nd Lumbricales	Latissimus dorsi Pectoralis major Extensor digitorum Extensor digiti minimi Extensor carpi ulnaris Extensor indicis Extensor pollicis longus
C7	**All of column 1 plus** Longus colli Serratus anterior Coracobrachialis Extensor carpi radialis, longus & brevis Abductor pollicis longus Extensor pollicis brevis Pronator teres Flexor carpi radialis Palmaris longus Abductor pollicis brevis Opponens pollicis Flexor pollicis brevis 1st & 2nd lumbricales	Neck extensors Scaleni Latissimus dorsi Pectoralis major Triceps Anconeus Extensor digitorum Extensor digiti minimi Extensor carpi ulnaris Extensor pollicis longus Extensor indicis	Flexor digitorum superficialis

C8	All of column 1 plus		
		Neck extensors	
		Pectoralis major	Pectoralis minor
	Scaleni	Pectoralis minor	
	Latissimus dorsi	Flexor digitorum	
	Triceps	superficialis	
	Anconeus	Flexor digitorum	
	Extensor digitorum	profundus, 1 & 2	
	Extensor digiti minimi	Flexor pollicis longus	
	Extensor carpi ulnaris	Pronator quadratus	
	Extensor pollicis longus	Flexor carpi ulnaris	
	Extensor indicis	Flexor digitorum	
	3rd & 4th lumbricales	profundus, 3 & 4	
		Abductor digiti minimi	
		Opponens digiti minimi	
		Flexor digiti minimi	
		Palmar interossei	
		Dorsal interossei	
		Adductor pollicis	
		Flexor pollicis brevis	

Table 2.5 Motor function in thoracic cord lesion

LEVEL	1. COMPLETE CONTROL	2. PARTIAL CONTROL	3. MINIMAL CONTROL
T1–4	**All upper limb muscles** Neck extensors		
			Trunk extensors
T5–6	**All upper limb muscles**		
			Trunk extensors
T7–8	**All upper limb muscles**		
			Trunk extensors
			Obliquus abdominis, internus & externus
			Rectus abdominis
			Transverse abdominus
T9–11	**All upper limb muscles**		
		Trunk extensors	
		Obliquus abdominis, internus & externus	
		Rectus abdominis	
		Transversus abdominis	
T12	**All upper limb muscles**		
	Obliquus abdominis externus	Trunk extensors	
	Rectus abdominis	Obliquus abdominis internus	
		Transversus abdominis	
		Quadratus lumborum	

Clinical guidelines

Table 2.6 Motor function in lumbosacral cord lesions

LEVEL	1. COMPLETE CONTROL	2. PARTIAL CONTROL	3. MINIMAL CONTROL
L1	**All the abdominals**		
	Quadratus lumborum Psoas minor Quadratus lumborum Psoas minor	Trunk extensors	
L2	**All of column 1**		
		Trunk extensors Psoas major Iliacus Sartorius	Pectineus Quadriceps
L3	**All column 1 plus**		
	Psoas major Iliacus Sartorius	Trunk extensors Pectineus Quadriceps Adductor brevis Adductor longus Gracilis Obturator externus Adductor magnus	
L4	**All of column 1 plus**		
	Pectineus Quadriceps Adductor brevis Adductor longus Gracilis Obturator externus Adductor magnus	Trunk extensors Flexor digitorum brevis Abductor hallucis 1st lumbrical	Gluteus medius Gluteus minimus Tensor fasciae latae Gemellus inferior Quadratus femoris Tibialis anterior Extensor hallucis longus Extensor digitorum longus Peroneus tertius Peroneus longus Peroneus brevis Plantaris Popliteus Flexor hallucis brevis
L5	**All of column 1 plus**		
	Flexor digitorum brevis Abductor hallucis 1st lumbrical	Trunk extensors Gluteus medius Gluteus minimus Tensor fascia latae Gemellus inferior Quadratus femoris Tibialis anterior Extensor hallucis longus Extensor digitorum, longus & brevis Peroneus tertius Peroneus longus Peroneus brevis Plantaris Popliteus Tibialis posterior Flexor digitorum longus Flexor hallucis brevis	Gluteus maximus Gemellus superior Obturator internus Semitendinosis Semimembranosis Biceps femoris (short head) Flexor hallucis longus

S1 **All of column 1 plus**

Gluteus medius	Trunk extensors	Biceps femoris (long head)
Gluteus minimus	Gluteus maximus	
Tensor fasciae latae	Piriformis	
Gemellus inferior	Gemellus superior	
Quadratus femoris	Obturator internus	
Tibialis anterior	Semitendinous	
Extensor hallucis longus	Semimembranosis	
Extensor digitorum longus	Biceps femoris (short head)	
Peroneus tertius	Gastrocnemius	
Extensor digitorum brevis	Soleus	
Peroneus longus	Flexor hallucis longus	
Peroneus brevis	Abductor digiti minimi	
Plantaris	Adductor hallucis	
Popliteus	Plantar interossei	
Tibialis posterior	Dorsal interossei	
Flexor digitorum longus	2nd, 3rd & 4th lumbricales	
Flexor hallucis brevis		

S2 **All of column 1 plus**

Gluteus maximus	Trunk extensors	Pelvic floor muscles
Piriformis	Biceps femoris (long head)	
Gemellus superior		
Obturator internus		
Semitendinosis		
Semimembranosis		
Biceps femoris (short head)		
Gastrocnemius		
Soleus		
Flexor hallucis longus		
Abductor digiti minimi		
Adductor hallucis		
Plantar interossei		
Dorsal interossei		
2nd, 3rd & 4th lumbricales		

S3 **All of column 1 plus**

Trunk extensors	Pelvic floor muscles

S4 **All of column 1 plus**

Pelvic floor muscles

Muscle tests

Key

In the subsequent descriptions of the method of manual muscle testing, the following symbols are used:

In the muscle lists preceding groups of tests –

Bold type = **prime mover**

Light type = secondary muscle action

In the photographs –

⇪   Direction of resistance

↑   Direction of movement

↕   Direction of muscle fibres

3. Evaluating the muscles of the upper extremity

Muscles tested at the scapula (Figs 3.1 & 3.2)

ELEVATION OF THE SCAPULA
  Levator scapula (LS)
  Trapezius — upper fibres (Tuf)

RETRACTION OF THE SCAPULA
  Rhomboideus major } **(Rh)**
  Rhomboideus minor }
  Trapezius — middle fibres (Tmf)

DEPRESSION OF THE SCAPULA
  Trapezius — lower fibres (Tlf)
  Subscapularis

PROTRACTION OF THE SCAPULA
  Serratus anterior (SA)
  Pectoralis minor
  Subclavius

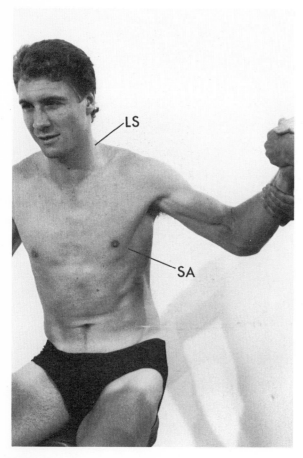

Fig. 3.1

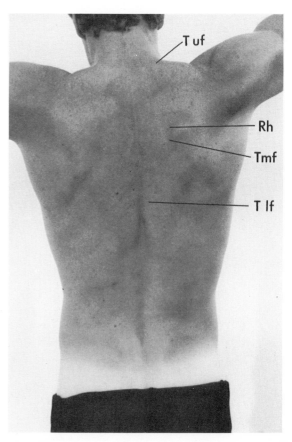

Fig. 3.2

LEVATOR SCAPULAE

Attachments: Extends from the posterior tubercles of the transverse processes of the first four cervical vertebrae to the medial border of the scapula, level with and above its spine.
Nerve supply: C3, 4, 5.

Surface markings: It can be palpated on the lateral surface of the neck immediately in front of the superior border of the upper fibres of trapezius.

Actions

Rostral end fixed: It elevates the scapula
Caudal end fixed: It assists in lateral flexion and rotation of the neck to the same side.

Maximum extensibility: The scapula is depressed, and the head rotated and side flexed to the opposite side.

TESTING POSITION

Grades 5, 4 and 3 (Fig. 3.3): In sitting, with the arms dependent and the head in the mid-position, the shoulder is elevated fully towards the ears. Resistance is applied to the upper surface of the shoulder in the direction of scapula depression.

Grades 2, 1 and 0 (Fig. 3.4): In supine lying, with the head in the mid position, the shoulder is elevated towards the ear.

Comments: It is advisable to test the right and left sides simultaneously, as any unilateral weakness will be detected more easily.

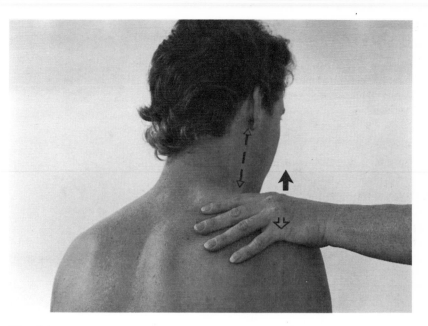

Fig. 3.3

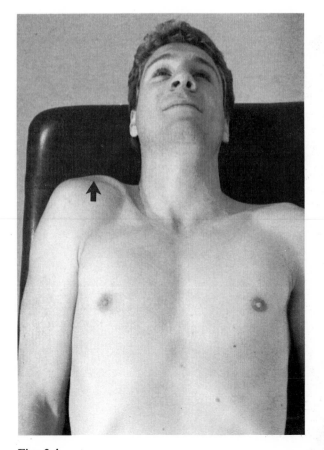

Fig. 3.4

Muscles tested at the scapula

TRAPEZIUS — UPPER FIBRES

Attachments: Extend from the external occipital protuberance, the medial third of the nuchal line, the ligamentum nuchae, and C7 spinous process to the lateral third of the clavicle.
Nerve supply: Accessory C3, 4.

Surface markings: The superior border of the muscle forms the posterior curve of the neck and can be palpated easily between its attachments.

Actions
Rostral end fixed: It elevates the scapula.
Caudal end fixed: Working bilaterally, it extends the neck. Working unilaterally, it rotates and laterally flexes the neck.

Maximum extensibility: The neck is flexed and rotated to the opposite side while the scapula is depressed.

TESTING POSITION
Grades 5, 4 and 3 (Fig. 3.5): In sitting, with the head turned towards the opposite side and stabilized, the shoulder is elevated towards the occiput. Resistance is applied to the upper surface of the shoulder in the direction of scapula depression.
Grades 2, 1 and 0 (Fig. 3.6): In prone lying, with the head rotated to the opposite side, the shoulder is elevated towards the occiput.

Comments: When tested bilaterally, the head will be in the midline. Thus, both levator scapulae and trapezius will be involved in the movement.

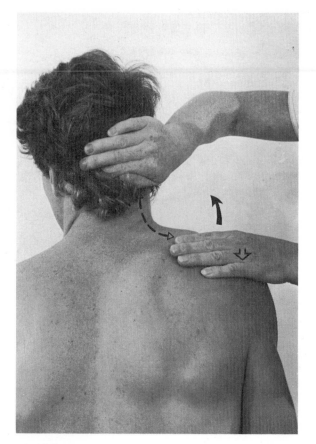

Fig. 3.5

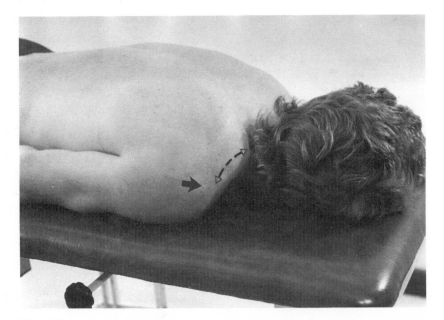

Fig. 3.6

TRAPEZIUS — MIDDLE FIBRES

Attachments: Extend from the ligamentum nuchae and the spinous processes of T1 – T5 to the medial margin of the acromion and the posterior superior surface of the spine of the scapula.

Nerve supply: Accessory, C3, 4.

Surface markings: The muscle is superficial and can be palpated between its attachments.

Actions

Rostral end fixed: It adducts the scapula.
Caudal end fixed: It will fixate the abducted scapula during upper limb weight bearing in activities such as push-ups.

Maximum extensibility: The scapula is abducted.

TESTING POSITION

Grades 5, 4 and 3 (Fig. 3.7): In prone lying, with the arm by the side of the trunk, the scapula is adducted. Resistance is applied to the posterior surface of the shoulder in the direction of scapular abduction.

Grades 2, 1 and 0 (Fig. 3.8): In sitting, with the arm at the side, the scapula is adducted.

Comments: When testing Grade 5, it may be necessary to increase the mechanical advantage of the operator. This can be done by abduction of the shoulder joint to 90°, allowing the forearm to hang vertically, and then applying resistance to the posterior surface of the elbow in the direction of horizontal flexion.

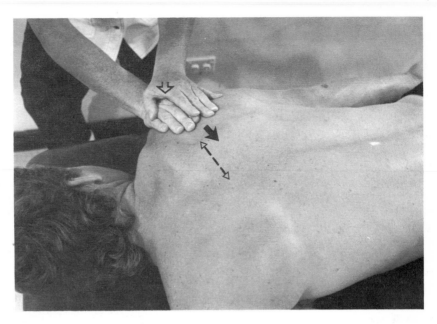

Fig. 3.7

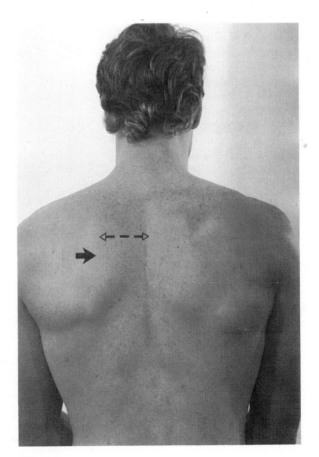

Fig. 3.8

Upper extremity

Muscles tested at the scapula

RHOMBOIDEUS, MAJOR AND MINOR

Attachments: Extend from the spinous processes and supraspinous ligaments of C7 – T5 to the medial border of the scapula between the spine and the inferior angle. Rhomboideus minor is the more rostral of the two.
Nerve supply: Dorsal scapular C4, 5.

Surface markings: These muscles can be palpated between their attachments, however, it is often difficult to differentiate between the rhomboids and middle fibres of trapezius.

Actions
Rostral end fixed: They adduct and elevate the scapula.
Caudal end fixed: The muscles stabilize the abducted scapula during weight bearing on the upper limbs, as in push-ups or crutch walking.

Maximum extensibility: The scapula is abducted and depressed.

TESTING POSITION
Grades 5, 4, and 3 (Fig. 3.9): In prone lying, with the shoulder in full internal rotation (the hand behind the back resting on the sacrum), the scapula is adducted and elevated. Resistance is applied to the posterior surface of the shoulder in the direction of scapular abduction and depression.
Grades 2, 1 and 0 (Fig. 3.10): In sitting, with the shoulder in full medial rotation (the hand behind the back resting on the sacrum), the scapula is adducted.

Comments: It is not possible to determine accurately the strength of these muscles in isolation from middle fibres of trapezius, as trapezius covers the rhomboids. However, in some spinal cord lesions where only some cervical nerve roots may have been preserved, this differentiating test may be useful. Similarly, it is not sensible to attempt to isolate the action of rhomboideus major from rhomboideus minor.

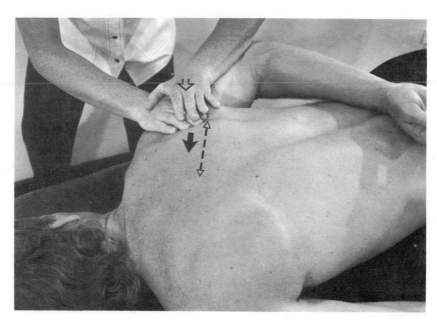

Fig. 3.9

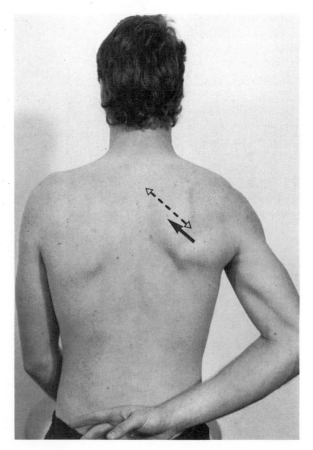

Fig. 3.10

TRAPEZIUS — LOWER FIBRES

Attachments: Extend from the spinous processes of the lower eight thoracic vertebrae and the associated supraspinal ligaments to the tubercle of the crest of the scapula.
Nerve supply: Accessory, C3, 4.

Surface markings: The muscle is superficial and can be palpated between its attachments.

Actions
Rostral end fixed: It depresses and adducts the scapula.
Caudal end fixed: It stabilizes the scapula during weight bearing on the upper limbs, as in crutch walking.

Maximum extensibility: The scapula is abducted and elevated with the inferior angle rotated upwards. To achieve this position it is easiest to use the upper limb as a lever and place the shoulder in full flexion with lateral rotation.

TESTING POSITION
Grades 5, 4 and 3 (Fig. 3.11): In prone lying, with the arm by the side of the trunk, the scapula is depressed and adducted. Resistance is applied to the posterior surface of the shoulder in the direction of elevation and abduction of the scapula.
Grades 2, 1 and 0 (Fig. 3.12): In prone lying, with the arm by the side of the trunk, the scapula is depressed and adducted.

Comments: The test position for Grades 5, 4 and 3 is not against gravity as that position is difficult for the patient to assume.

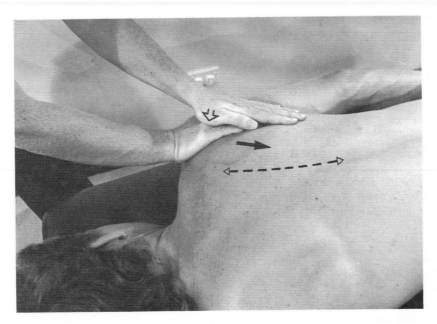

Fig. 3.11

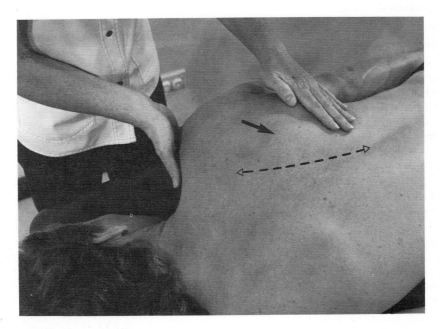

Fig. 3.12

Muscles tested at the scapula

SERRATUS ANTERIOR

Attachments: Extends from the superior borders and outer surfaces of the first eight or nine ribs (close to their angles) to the medial border of the costal surface of the scapula.

Nerve supply: Long thoracic C5, 6, 7.

Surface markings: The muscle lies on the lateral surface of the thorax, congruent with the inferior border of pectoralis major. The interdigitations of the muscle between the ribs can be palpated.

Actions

Rostral end fixed: It protracts and rotates the inferior angle of the scapula forwards.

Caudal end fixed: It stabilizes the scapula on the thorax during weight bearing on the upper limbs, as in push-ups or chin-ups.

Maximum extensibility: The scapula is fully adducted with the inferior angle of the scapula rotated towards the vertebral column.

TESTING POSITION

Grades 5, 4 and 3 (Fig. 3.13): In supine lying, with the shoulder flexed to 90° and the elbow fully flexed, the shoulder complex is protracted.

Grades 2, 1 and 0 (Fig. 3.14): In sitting, with the upper arm supported in 90° of shoulder flexion and the elbow fully flexed, the shoulder complex is protracted.

Comments: Winging of the scapula indicates weakness of this muscle.

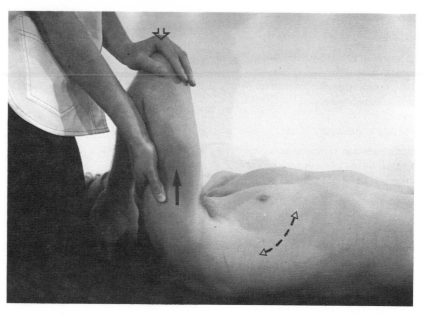

Fig. 3.13

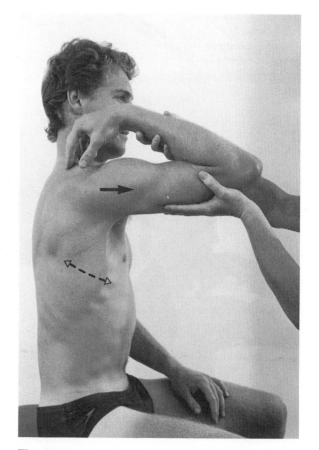

Fig. 3.14

PECTORALIS MINOR

Attachments: Extends from the superior margins of the third, fourth and fifth ribs near the costal cartilages, to the superior surface and medial border of the coracoid process.
Nerve supply: Medial pectoral C7, 8, T1.

Surface markings: It can be palpated just below the coracoid process through a relaxed pectoralis major.

Actions
Rostral end fixed: It moves scapula forward and caudally.
Caudal end fixed: It assists with forced expiration.

Maximum extensibility: Full inspiration with the scapula held in adduction.

TESTING POSITION
Grades 5, 4 and 3 (Not illustrated): In supine, with the arms by the side of the trunk, the patient lifts the shoulder away from the supporting surface. Resistance is applied on the anterior surface of the shoulder in a vertical direction towards the supporting surface.
Grades 2, 1 and 0 (Not illustrated): In sitting, with the arms by the side of the trunk, the subject moves the shoulder forwards.

Comments: The humerus must not be used as a lever to force the shoulder forwards. This muscle is usually tested in conjunction with pectoralis major.

SUBCLAVIUS

Attachments: Extends from the costochondral junction of the first rib to the inferior surface of the clavicle.
Nerve supply: C5, 6.

Actions
Rostral end fixed: Its function is obscure but it may be involved in depressing the lateral portion of the clavicle.
Caudal end fixed: Not known.

Maximum extensibility: Elevation and adduction of the shoulder girdle.

TESTING POSITION
Since the muscle's action is obscure, strength testing is inappropriate.

Comments: The muscle has been described here only for completeness.

Upper extremity

Muscles tested at the shoulder joint (Figs 3.15 & 3.16)

FLEXION OF THE SHOULDER
JOINT
    Coracobrachialis (Cb)
      Pectoralis major (PMj)
    Deltoid — anterior fibres

EXTENSION OF THE SHOULDER
JOINT
    Latissimus dorsi (LD)
    Deltoid — posterior fibres (D)

ABDUCTION OF THE SHOULDER
JOINT
    Deltoid — middle fibres
    Supraspinatus

ADDUCTION OF THE SHOULDER
JOINT
    Pectoralis major (PMj)

EXTERNAL ROTATION OF THE
SHOULDER JOINT
    Teres minor (TMn)
    Infraspinatus (Is)

INTERNAL ROTATION OF THE
SHOULDER JOINT
    Teres major (TMj)
    Subscapularis

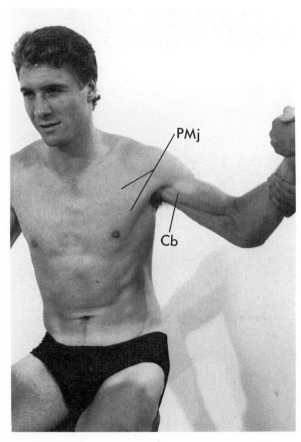

Fig. 3.15

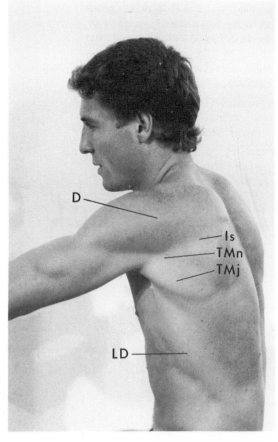

Fig. 3.16

CORACOBRACHIALIS

Attachments: Extends from the coracoid process of the scapula to the medial border of the middle third of the humerus.

Nerve supply: Musculocutaneous C5, 6, 7.

Surface markings: Its tendon can be palpated near the anterior border of the axilla, adjacent to the upper medial border of biceps brachii.

Actions

Rostral end fixed: It flexes and adducts the humerus.

Caudal end fixed: It assists deltoid in stabilizing the gleno-humeral joint in weightbearing through the upper limb, as in crutch walking.

Maximum extensibility: The humerus is fully extended and then abducted.

TESTING POSITION

Grades 5, 4 and 3 (Fig. 3.17): In sitting, with the shoulder joint slightly abducted and the elbow supported in 90° of flexion, the upper arm is flexed and adducted across the trunk. Resistance is applied to the medial surface of the elbow joint in the direction of extension and abduction.

Grades 2 and 1 (Fig. 3.18): In side lying, with the upper arm supported, the elbow flexed and forearm supinated, the upper arm is flexed and adducted across the trunk.

Comments: It is difficult to differentiate the action of this muscle from that of the anterior fibres of deltoid. It should always be palpated to ascertain its contribution to the movement.

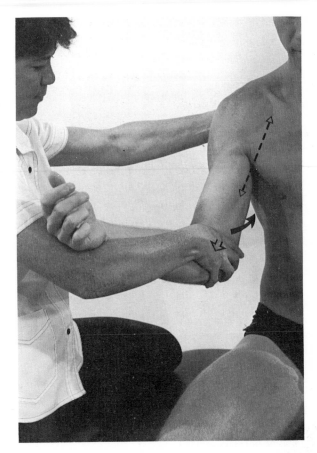

Fig. 3.17

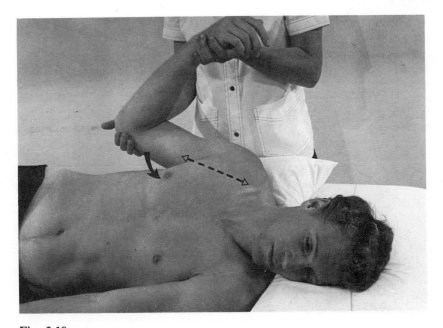

Fig. 3.18

Upper extremity

Muscles tested at the shoulder joint

DELTOID

Attachments: The anterior fibres extend from the lateral third of the clavicle, the middle fibres from the lateral border of the acromion process and the posterior fibres from the lower border of the spine of the scapula. All fibres attach to the deltoid tuberosity of the humerus.
Nerve supply: Axillary C5, 6.

Surface markings: It forms the rounded elevation of the shoulder. The anterior fibres can be palpated on the anterior aspect of the shoulder, the middle fibres on the lateral aspect of the shoulder, and the posterior fibres on the posterior aspect of the shoulder.

Actions
Rostral end fixed: The anterior fibres assist in shoulder flexion and internal rotation. The posterior fibres assist in shoulder extension and external rotation. The middle fibres abduct the shoulder.
Caudal end fixed: Assists in the stabilization of the shoulder joint during upper limb weight bearing.

Maximum extensibility: For the posterior and middle fibres the arm is horizontally adducted with the humerus internally rotated. For the anterior and middle fibres the arm is adducted behind the trunk with the humerus externally rotated.

TESTING POSITION — Anterior fibres
Grades 5, 4 and 3 (Fig. 3.19): In sitting, the elbow in 90° of flexion, the humerus is flexed. Resistance is applied to the anterior surface of the upper arm, just above the elbow in the direction of extension.
Grades 2, 1 and 0 (Fig. 3.20): In side lying, with the arm to be tested uppermost and supported and the elbow in 90° of flexion, the humerus is flexed.

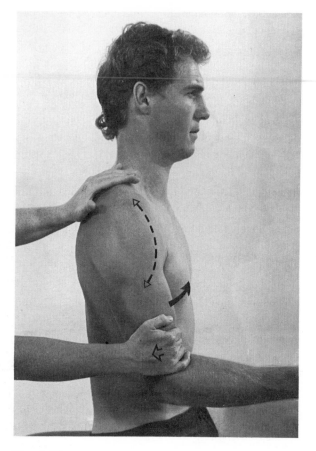

Fig. 3.19

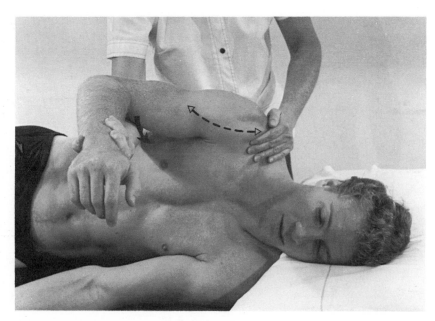

Fig. 3.20

LATISSIMUS DORSI

Attachments: Extends from the spinous processes and supraspinous ligaments of the seventh to twelfth thoracic and all the lumbar vertebrae, the median ridge of the sacrum, the posterior part of the iliac crest, the lower three ribs, and the inferior angle of the scapula, to the floor of the bicipital groove of the humerus.
Nerve supply: Thoracodorsal C6, 7, 8.

Surface markings: As the muscle is superficial it is palpated easily between its attachments.

Actions
Rostral end fixed: Bilaterally, it assists in stabilizing the pelvis when walking with an aid such as crutches or a stick, and in lifting the pelvis during chair to chair transfers in the paraplegic person.
Caudal end fixed: It internally rotates, adducts and extends the humerus. It may also assist in lateral flexion of the trunk, and extension of the lower trunk.

Maximum extensibility: Full flexion, abduction and external rotation of the shoulder. There should be no increase in curvature of the lumbar or thoracic spine.

TESTING POSITION
Grades 5, 4 and 3 (Fig. 3.21): To test the middle and inner ranges the subject is placed in prone lying and the humerus is extended, adducted and internally rotated. Resistance is applied to the medial surface of the humerus at the elbow joint, in the direction of flexion, abduction and external rotation. To test the outer and middle ranges, (not illustrated) the subject is placed in supine lying with the arm fully flexed. The arm is extended, adducted and internally rotated from this elevated and slightly abducted position above the head. Resistance is applied to the medial surface of the humerus in the direction of flexion abduction and external rotation.
Grades 2, 1 and 0 (Fig. 3.22): In side lying, with the limb to be tested uppermost, the upper arm is extended and internally rotated from the flexed position. The adduction component cannot be practically assessed in the gravity eliminated position.

Comments: In the last 10° of the

movement care must be taken to ensure that the action occurs at the gleno-humeral joint only. It is easy to substitute anterior depression of the scapula in this range.

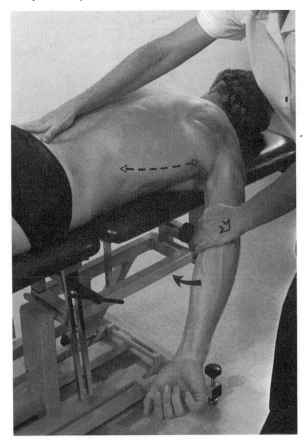

Fig. 3.21

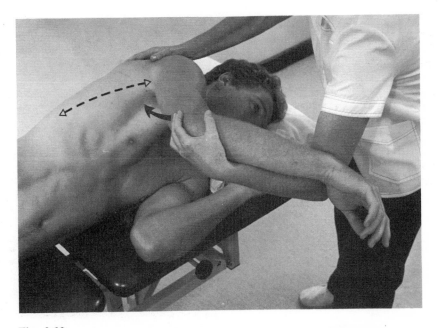

Fig. 3.22

Upper extremity

Muscles tested at the shoulder joint

DELTOID (see p. **40** for attachments, surface markings, actions, maximum extensibility)

TESTING POSITION — posterior fibres
Grades 5, 4 and 3 (Fig. 3.23): In sitting, the elbow in 90° of flexion, the humerus is extended. Resistance is applied to the posterior surface of the upper arm just above the elbow, in the direction of flexion.
Grades 2, 1 and 0 (Fig. 3.24): In side lying, with the arm to be tested uppermost and supported, the elbow in 90° of flexion, the humerus is extended.

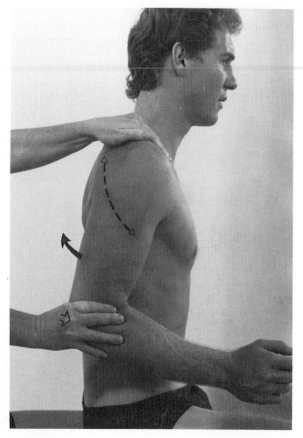

Fig. 3.23

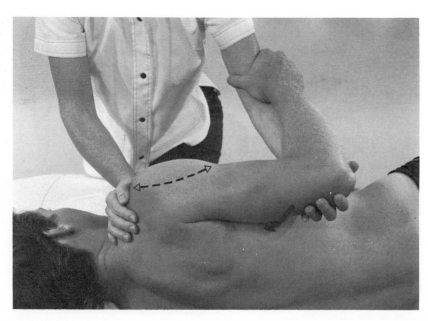

Fig. 3.24

DELTOID (see p. **40** for attachments, surface markings, actions, maximum extensibility)

TESTING POSITION — middle fibres
Grades 5, 4 and 3 (Fig. 3.25): In sitting, the humerus is abducted to 90°. Resistance is applied to the lateral surface of the upper arm just above the elbow in the direction of adduction.
Grades 2, 1 and 0 (Fig. 3.26): In supine lying, with the elbow in 90° of flexion, the supported humerus is abducted to 90°.

Comments: As abduction can be performed by biceps brachii if the humerus is allowed to externally rotate, the humerus must be held midway between internal and external rotation and the elbow in 90° of flexion to prevent biceps brachii from producing the movement.

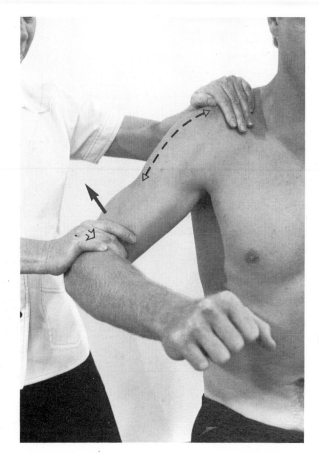

Fig. 3.25

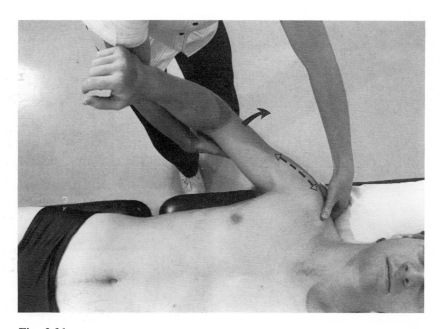

Fig. 3.26

Upper extremity

Muscles tested at the shoulder joint

SUPRASPINATUS

Attachments: Extends from the supraspinous fossa of the scapula to the greater tuberosity of the humerus, passing beneath the arch of the acromion process.
Nerve supply: Suprascapular C4, 5, 6.

Surface markings: The muscle belly can be palpated in the lateral part of the supraspinous fossa of the scapula.

Actions
Rostral end fixed: It holds the head of the humerus in the glenoid socket and assists deltoid in abducting the humerus.
Caudal end fixed: It assists in stabilizing the scapula during upper limb weight bearing, as in crutch walking.

Maximum extensibility: The humerus is adducted and internally rotated behind the trunk such that the forearm is parallel to the sacrum.

TESTING POSITION
Grades 5, 4 and 3 (Fig. 3.27): In sitting, the humerus is abducted 15°. Resistance is applied to the lateral surface of the humerus just above the elbow joint, in the direction of adduction.
Grades 2, 1 and 0 (Fig. 3.28): In supine lying, with the elbow in 90° of flexion, the supported humerus is abducted 15°.

Comments: The tendinous caudal attachment of the supraspinatus muscle is the most commonly damaged component of the musculotendinous cuff which surrounds the shoulder joint. Loss of its stabilising influence will disrupt the initiation of shoulder abduction.

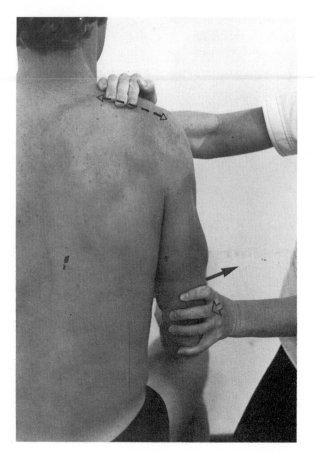

Fig. 3.27

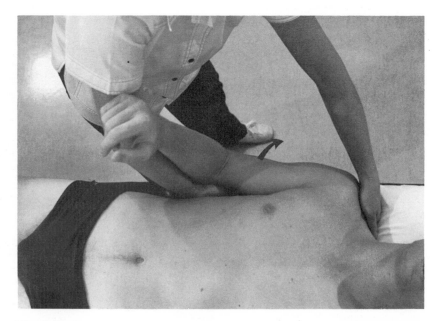

Fig. 3.28

PECTORALIS MAJOR

Attachments: Extends from the anterior surface of the medial half of the clavicle (clavicular head) and the anterior surface of the manubrium sternum, plus the costal cartilages of the first 6 or 7 ribs (sternal head) to the whole of the lateral lip of the bicipital groove.
Nerve supply: Lateral and medial pectoral C5, 6, 7, 8, T1.

Surface markings: It forms the anterior margin of the axilla.

Actions
Rostral end fixed: Both heads act together to adduct and medially rotate the humerus. The clavicular head flexes and medially rotates the humerus as well as horizontally adducting it. The sternal head also extends from a flexed position.
Caudal end fixed: It assists in stabilizing the shoulder girdle during upper limb weight bearing. It can assist also in inspiration.

Maximal extensibility: With the opposite shoulder stabilized by placing the arm behind the back, the shoulder to be tested is fully externally rotated and abducted maximally.

TESTING POSITION
Grades 5, 4 and 3

Clavicular head (not illustrated): In supine lying, with the shoulder abducted to 90°, the arm is horizontally adducted towards the opposite side. Resistance is applied to the anterior surface of the arm just above the elbow joint in the direction of horizontal abduction.

Sternal head (Fig. 3.29): In supine lying, with the shoulder flexed and slightly abducted above the head, the arm is moved towards the opposite hip. Resistance is applied to the anterior surface of the arm at the elbow joint, in the direction of flexion, abduction and external rotation.

Grades 2, 1 and 0 (Fig. 3.30)

Clavicular head: In sitting, with the shoulder abducted to 90°, the arm is horizontally adducted towards the opposite side.

Sternal head: It is not practical to test these fibres in a gravity eliminated position.

Comments: In either of the positions for testing Grades 5, 4 and 3, only the outer and middle ranges of the contraction are against gravity.

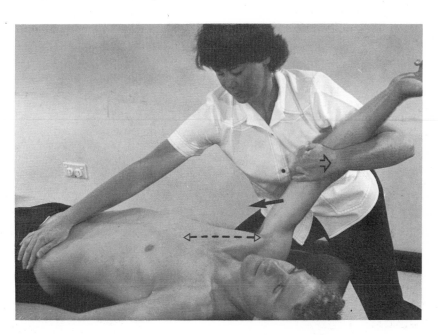

Fig. 3.29

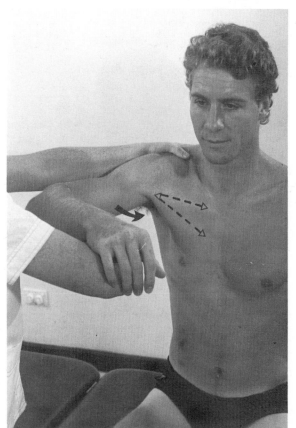

Fig. 3.30

Upper extremity

Muscles tested at the shoulder joint

TERES MINOR

Attachments: Extends from the upper two-thirds of the axillary border of the scapula to the lowest facet of the greater tuberosity of the humerus.
Nerve supply: Axillary C5, 6.

Surface markings: This muscle can be palpated between the posterior fibres of deltoid and teres major.

Actions
Rostral end fixed: It adducts and externally rotates the humerus.
Caudal end fixed: It assists in stabilizing the scapula during upper limb weight bearing, as in push-ups and crutch walking.

INFRASPINATUS

Attachments: Extends from the infraspinous fossa of the scapula to the greater tuberosity of the humerus.
Nerve supply: Suprascapular C5, 6.

Surface markings: The muscle can be palpated over the dorsal surface of the scapula below the spine.

Actions
Rostral end fixed: It extends and externally rotates the humerus.
Caudal end fixed: It assists in stabilizing the scapula during upper limb weight bearing, as in push-ups and crutch walking.

For teres minor and infraspinatus

Maximum extensibility: With the humerus in 90° of abduction, the shoulder joint is internally rotated.

TESTING POSITION
Grades 5, 4 and 3 (Fig. 3.31): In prone lying, with the supported humerus in 90° of abduction and the elbow at right angles, the shoulder joint is externally rotated by moving the forearm towards the ceiling. Only the middle and inner ranges are against gravity. Resistance is applied to the posterior surface of the wrist in the direction of internal rotation of the shoulder.
Grades 2, 1 and 0 (Fig. 3.32): In side lying, with the supported humerus in 90° of abduction and the elbow flexed to 90°, the shoulder joint is externally rotated by moving the forearm rostrally.

Comments: Functionally, these muscles cannot be satisfactorily differentiated and are therefore tested together.

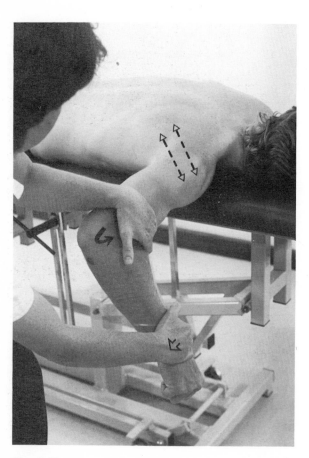

Fig. 3.31

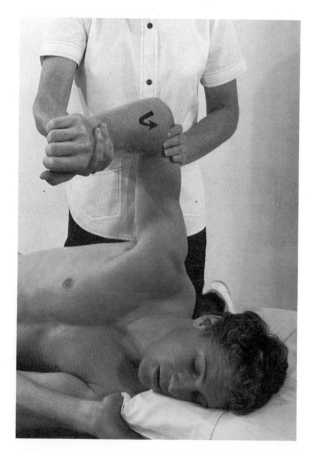

Fig. 3.32

TERES MAJOR

Attachments: Extends from the lower third of the posterior axillary border of the scapula, to the medial border of the bicipital groove of the humerus.
Nerve supply: Lower subscapular C6, 7.

Surface markings: It can be palpated between its attachments, and is often a very prominent muscle in males.

Actions

Rostral end fixed: It adducts, extends and internally rotates the humerus.
Caudal end fixed: It stabilizes the scapula during upper limb weight bearing, as in push-ups and crutch walking.

SUBSCAPULARIS

Attachments: Extends from the subscapular fossa to the lesser tuberosity of the humerus and the anterior part of the shoulder joint capsule.
Nerve supply: Upper and lower subscapular C5, 6.

Surface markings: It lies under the scapula and can only be palpated in the axilla close to its insertion.

Actions

Rostral end fixed: It internally rotates the humerus. It also stabilizes the head of the humerus in the glenoid cavity during movement.
Caudal end fixed: It stabilizes the scapula during upper limb weight bearing, as in push-ups and crutch walking.

For teres major and subscapularis

Maximum extensibility: With the humerus in 90° of abduction the shoulder joint is externally rotated.

TESTING POSITION

Grades 5, 4 and 3 (Fig. 3.33): In prone lying, with the shoulder in 90° of abduction and the elbow joint flexed to 90°, the humerus is internally rotated by moving the forearm towards the ceiling. Only the middle and inner ranges of this movement are against gravity. Resistance is applied to the anterior surface of the wrist in the direction of external rotation.
Grades 2, 1 and 0 (Fig. 3.34): In side lying, with the shoulder in 90° of abduction and the elbow flexed to 90° and supported, the humerus is internally rotated by moving the forearm caudally.

Comments: Functionally, these muscles cannot be satisfactorily differentiated and are therefore tested together. The outer range can be assessed against gravity in a supine position.

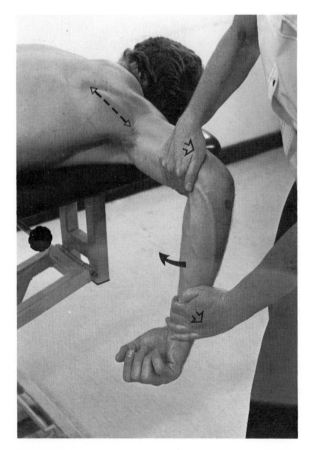

Fig. 3.33

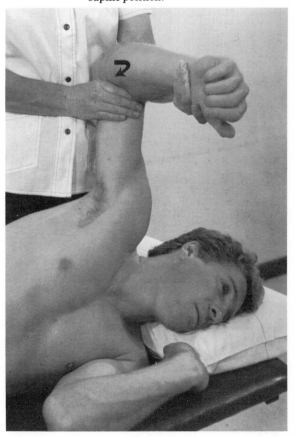

Fig. 3.34

Upper extremity

Muscles tested at the elbow joint (Figs 3.35 & 3.36)

FLEXION OF THE ELBOW JOINT
  Biceps brachii (Bb)
  Brachialis (Br)
  Brachioradialis (BR)

EXTENSION OF THE ELBOW
JOINT
  Anconeus
  Triceps (T)

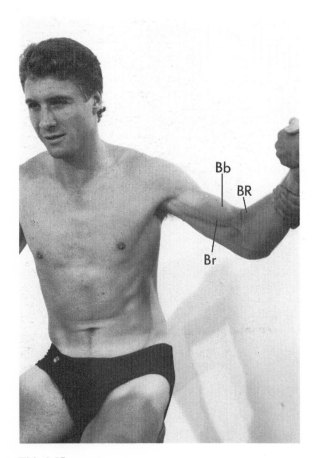

Fig. 3.35

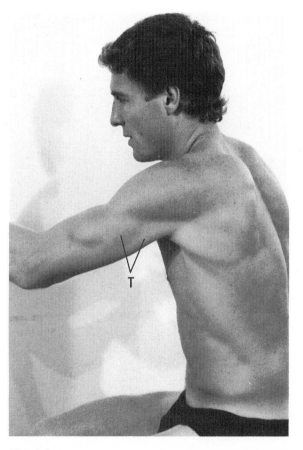

Fig. 3.36

BICEPS BRACHII

Attachments: The long head originates from the supraglenoid tubercle of the scapula and the short head from the coracoid process of the scapula. Both heads insert into the bicipital tuberosity of the radius.

Nerve supply: Musculocutaneous, C5, 6.

Surface markings: The muscle belly forms the elevation on the anterior surface of the arm which is delineated by flexing the supinated forearm against resistance. The biceps tendon can be palpated in the cubital fossa, where it is the most superficial midline tendon.

Actions

Rostral end fixed: It flexes the elbow joint such that the forearm moves towards the arm, and supinates the forearm. Also, it assists in flexing and abducting the shoulder joint (if the humerus is externally rotated).

Caudal end fixed: It flexes the elbow joint such that the arm is flexed towards the forearm, as in chin-ups.

Maximum extensibility: The shoulder joint is extended fully and externally rotated by the side of the trunk. The elbow must be extended and the forearm pronated.

TESTING POSITION

Grades 5, 4 and 3 (Fig. 3.37): In sitting, with the arm by the side of the trunk, the supinated forearm is flexed. Resistance is applied to the anterior surface of the wrist in the direction of elbow extension.

Grades 2, 1 and 0 (Fig. 3.38): In sitting, with the arm supported so that the shoulder is abducted to 90°, the supinated forearm is flexed.

Comments: This muscle can be tested in supine for Grades 5, 4 and 3 if the subject cannot sit, however the inner range is not tested against gravity in that position. Grades 2, 1 and 0 can also be tested in side lying.

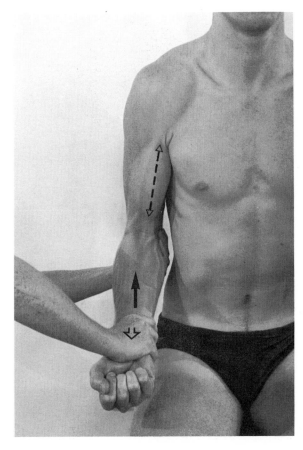

Fig. 3.37

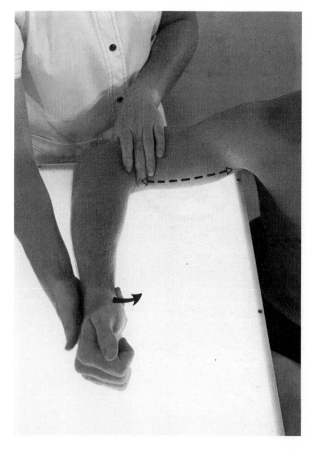

Fig. 3.38

Upper extremity

Muscles tested at the elbow joint

BRACHIALIS

Attachments: Extends from the lower two-thirds of the anterior surface of the humerus to the coronoid process of the ulna.

Nerve supply: Musculocutaneous (C5, 6) and a small branch from the radial (C7).

Surface markings: It is deep to biceps brachii. Its tendon can be palpated medial and deep to that of the biceps brachii in the cubital fossa.

Actions

Rostral end fixed: It flexes the elbow joint.
Caudal end fixed: It flexes the elbow joint such that the arm is flexed towards the forearm, as in chin-ups.

Maximum extensibility: Extension of the elbow with full supination of the forearm.

TESTING POSITION

Grades 5, 4 and 3 (Fig. 3.39): In sitting, with the arm by the side of the trunk, the pronated forearm is flexed. Resistance is applied to the posterior surface of the wrist in the direction of elbow extension.

Grades 2, 1 and 0 (Fig. 3.40): In sitting, with the arm supported so that the shoulder is abducted to 90°, the pronated forearm is flexed.

Comments: This muscle can also be tested in supine for Grades 5, 4 and 3, however the inner range is not tested against gravity in that position. Grades 2, 1 and 0 can also be tested in side lying.

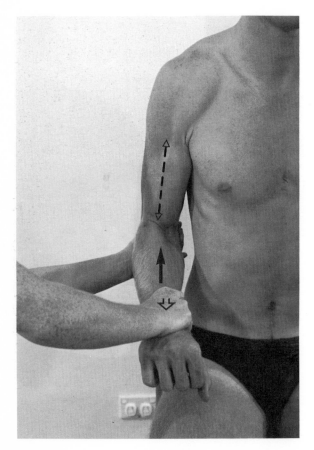

Fig. 3.39

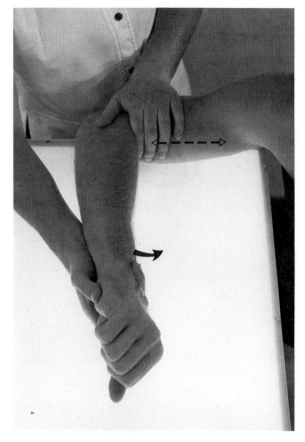

Fig. 3.40

BRACHIORADIALIS

Attachments: Extends from the lateral supracondylar ridge of the humerus to the base of the styloid process of the radius.

Nerve supply: Radial C5, 6.

Surface markings: It can be palpated on the lateral aspect of the forearm.

Actions

Rostral end fixed: With the forearm in a mid-position, it flexes the elbow joint. It can assist in pronation and also in supination.

Caudal end fixed: It flexes the elbow joint such that the arm is flexed towards the forearm, as in chin-ups.

Maximum extensibility: With the forearm pronated, the elbow is extended.

TESTING POSITION

Grades 5, 4 and 3 (Fig. 3.41): In sitting, with the arm by the side of the trunk and the forearm held in mid-position, the elbow is flexed. Resistance is applied to the radial surface of the wrist in the direction of elbow extension.

Grades 2, 1 and 0 (Fig. 3.42): In sitting, with the arm supported so that the shoulder is abducted to 90° and the forearm held in mid-position, the elbow joint is flexed.

Comments: This muscle can be tested in supine for Grades 5, 4 and 3 if the subject cannot sit, however the inner range is not tested against gravity in that position. Grades 2, 1 and 0 can also be tested in side lying.

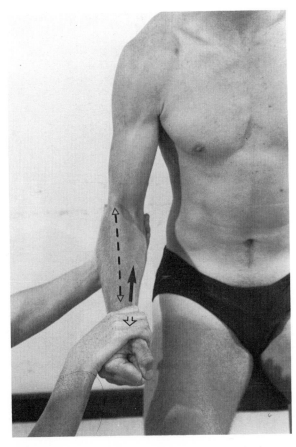

Fig. 3.41

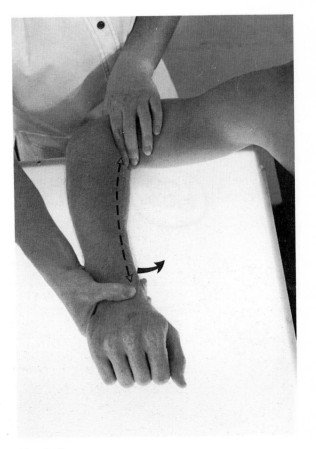

Fig. 3.42

Upper extremity

Muscles tested at the elbow joint

ANCONEUS

Attachments: Extends from the posterior surface of the lateral epicondyle of the humerus to the olecranon process and posterior surface of the ulna.
Nerve supply: Radial C7, 8.

TRICEPS

Attachments: The long head originates from the axillary border of the scapular below the glenoid fossa, the lateral head from the posterior and lateral surface of the humerus below the greater tuberosity and the medial head from the posterior surface of the humerus below the radial groove. All heads insert onto the posterior surface of the olecranon process of the ulna.
Nerve supply: Radial C6, 7, 8.

For anconeus and triceps

Surface markings: Both of these muscles are superficial and can be palpated between their attachments.

Actions
Rostral end fixed: These muscles extend the elbow joint. The long head of triceps also assists in adduction and extension of the shoulder joint.
Caudal end fixed: These muscles extend the elbow when weight bearing on the upper extremity, as in push-ups.

Maximum extensibility: With the shoulder externally rotated and flexed, the elbow joint is flexed.

TESTING POSITION
Grades 5, 4 and 3 (Fig. 3.43): In supine lying, with the shoulder joint flexed to 90°, the elbow joint is extended.

Grades 2, 1 and 0 (Fig. 3.44): In sitting, with the arm supported so that the shoulder is abducted to 90° and the forearm is in a mid-position, the elbow joint is extended.

Comments: Grades 5, 4 and 3 for triceps can also be tested in prone lying with the shoulder abducted to 90°. It will, however, appear weaker than when tested as described above. This is because the prone position puts the long head of triceps in a shortened position over both the shoulder and elbow joints.

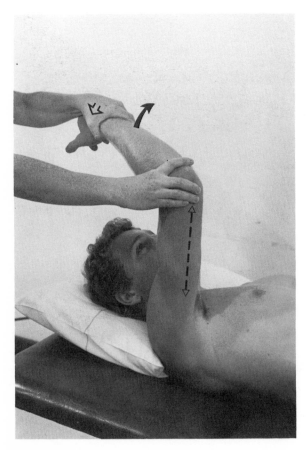

Fig. 3.43

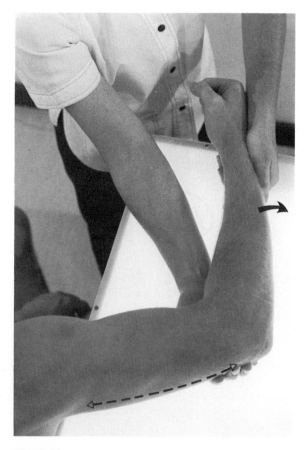

Fig. 3.44

Muscles tested at the superior and inferior radio-ulnar joints (Figs 3.45 & 3.46)

SUPINATION OF THE RADIO-
ULNAR JOINTS
Supinator (S)
Biceps brachii

PRONATION OF THE RADIO-
ULNAR JOINTS
Pronator teres (PT)
Pronator quadratus (PQ)

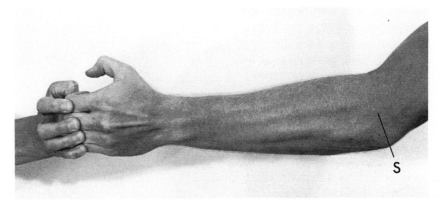

Fig. 3.45

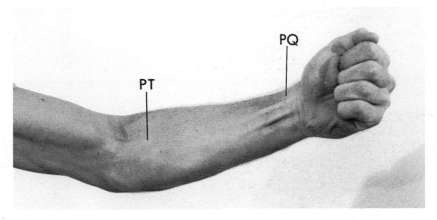

Fig. 3.46

Upper extremity

Muscles tested at the superior and inferior radio-ulnar joints

SUPINATOR

Attachments: Extends from the lateral condyle of the humerus and the supinator crest of the ulna, to wrap around the anterior and lateral surface of the upper third of the radius.
Nerve supply: Posterior interosseous C5, 6.

Surface markings: With the wrist extensors relaxed, the muscle can be palpated on the posterior surface of the forearm behind to the head of the radius.

Actions
Rostral end fixed: It supinates the forearm.
Caudal end fixed: It assists in stabilizing the extended elbow during upper limb weight bearing, such as in the 'L' support and straight support positions on the Roman Rings.

Maximum extensibility: With the elbow extended, the forearm is pronated.

TESTING POSITION
Grades 5, 4 and 3 (Fig. 3.47): In sitting, with the arm by the side of the trunk and the elbow flexed to 90°, the supported forearm is supinated. Only the outer and middle ranges are against gravity. Resistance is applied to the posterior surface of the wrist in the direction of pronation.
Grades 2, 1 and 0 (Fig. 3.48): With the supported humerus in abduction and medial rotation and the elbow joint flexed to 90°, the forearm is supinated.

Comments: To reduce the amount of involvement of biceps, supination of the forearm can be performed with the elbow fully flexed (with the shoulder in some flexion) or extended (with the shoulder in some extension) as these positions put biceps in either a shortened or lengthened position and reduce its efficiency.

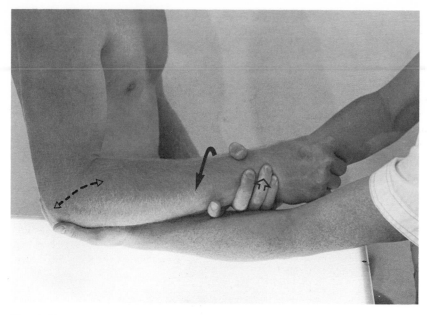

Fig. 3.47

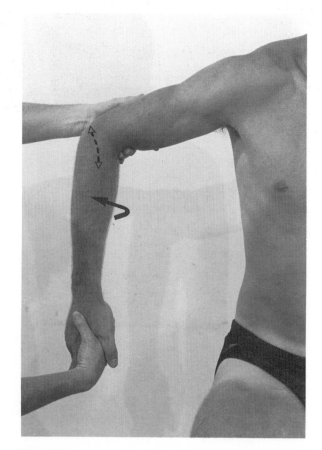

Fig. 3.48

Muscles tested at the superior and inferior radio-ulnar joints

PRONATOR TERES

Attachments: The superficial head originates from the medial condyle of the humerus and the deep head from the medial side of the coracoid process of the ulna. Both heads insert onto the middle third of the lateral surface of the radius.
Nerve supply: Median C6, 7.

Surface markings: It forms the medial border of the cubital fossa.

Actions

Rostral end fixed: It pronates the forearm and also assists with flexion of the elbow.
Caudal end fixed: It assists in stabilizing the extended elbow during upper limb weight bearing, such as in the 'L' support and straight support positions on the Roman Rings.

PRONATOR QUADRATUS

Attachments: It extends from the lower quarter of the anterior surface of the ulna to the lower quarter of the anterior surface of the radius.
Nerve supply: Anterior interosseous C7, 8, T1.

Surface markings: As it is deep to the flexor tendons it is difficult to palpate.

Actions

Rostral end fixed: It pronates the forearm.
Caudal end fixed: Not appropriate to this muscle, as it runs transversely.

For pronator teres and pronator quadratus

Maximum extensibility: With the elbow extended, the forearm is supinated.

TESTING POSITION

Grades 5, 4 and 3 (Fig. 3.49): In sitting, with the arm by the side of the trunk and the elbow flexed to 90°, the supported forearm is pronated. Only the outer and middle ranges are against gravity. Resistance is applied to the anterior surface of the wrist in the direction of supination. This puts pronator teres in a shortened position and therefore makes it less effective but eliminates the effect of external rotation of the shoulder on this movement.

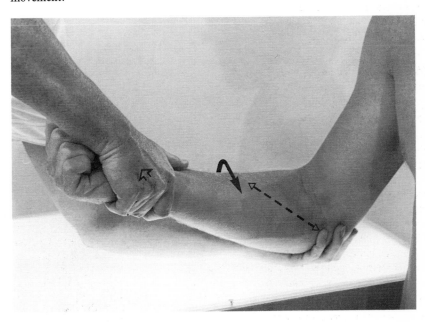

Fig. 3.49

Grades 2, 1 and 0 (Not illustrated): With the supported humerus in abduction and medial rotation and the elbow joint flexed to 90°, the forearm is pronated. Same position as in Figure 3.48, page **54**.

Comments: It is important to prevent any shoulder abduction or internal rotation during the testing procedure.

Upper extremity

Muscles tested at the wrist joint (Figs 3.50 & 3.51)

FLEXION OF THE WRIST IN THE
NEUTRAL POSITION
Palmaris longus (PL)
    Flexor carpi ulnaris
    Flexor carpi radialis

FLEXION OF THE WRIST WITH
ULNA DEVIATION
Flexor carpi ulnaris (FCU)

FLEXION OF THE WRIST WITH
RADIAL DEVIATION
Flexor carpi radialis (FCR)

EXTENSION OF THE WRIST IN
THE NEUTRAL POSITION
    Extensor carpi radialis longus
    Extensor carpi radialis brevis
    Extensor carpi ulnaris

EXTENSION OF THE WRIST WITH
RADIAL DEVIATION
**Extensor carpi radialis longus
(ECRL)**
**Extensor carpi radialis brevis
(ECRB)**

EXTENSION OF THE WRIST WITH
ULNA DEVIATION
Extensor carpi ulnaris (ECU)

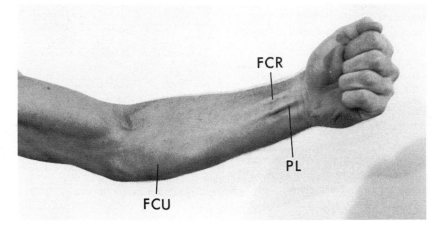

Fig. 3.50

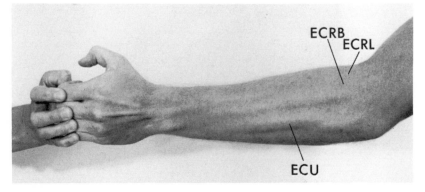

Fig. 3.51

PALMARIS LONGUS

Attachments: Extends from the medial epicondyle of the humerus to the flexor retinaculum and palmar aponeurosis.
Nerve supply: Median C7, 8.

Surface markings: The tendon can be palpated as the most superficial on the anterior surface of the wrist mid-way between the radius and ulna.

Actions
Rostral end fixed: It tenses the palmar fascia and flexes the wrist. It may also assist in elbow flexion.
Caudal end fixed: It stabilizes the wrist when weight bearing through the tips of the digits, as in finger-tip push-ups.

Maximum extensibility: With the forearm supinated, both the elbow and wrist joints are extended.

TESTING POSITION
Grades 5, 4 and 3 (Fig. 3.52): With the forearm supported in supination, the hand is cupped and the wrist flexed. Resistance is applied to the palm of the hand in the direction of wrist extension.
Grades 2, 1 and 0 (Fig. 3.53): With the forearm supported mid-way between supination and pronation, so that the ulnar surface of the forearm is resting on the support, the hand is cupped and the wrist flexed.

Comments: This muscle is not always present.

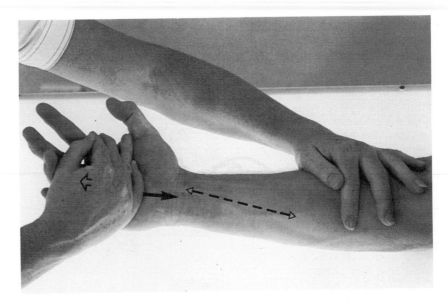

Fig. 3.52

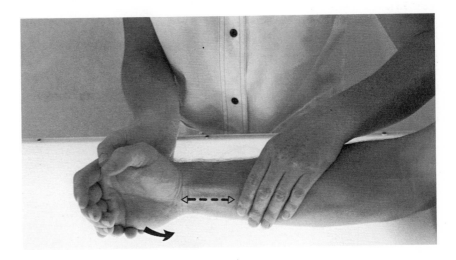

Fig. 3.53

Upper extremity

Muscles tested at the wrist joint

FLEXOR CARPI ULNARIS

Attachments: The humeral head originates from the medial epicondyle of the humerus and the ulnar head from the medial surface of the olecranon and the rostral two-thirds of the posterior border of the ulna. Both heads insert into the pisiform, hamate and fifth metacarpal.
Nerve supply: Ulnar C7, 8.

Surface markings: It is the most medial of the flexor tendons at the wrist. It can be palpated on the anterior surface of the wrist, as it crosses to insert onto the pisiform bone.

Actions
Rostral end fixed: It flexes and adducts the wrist joint. It may assist also in flexion of the elbow joint.
Caudal end fixed: It assists in stabilizing the wrist joint during weight bearing through the tips of the digits, as in finger-tip push-ups.

Maximum extensibility: With the elbow joint extended, the wrist joint is radially deviated and extended.

TESTING POSITION
Grades 5, 4 and 3 (Fig. 3.54): With the forearm supported in supination, the wrist is flexed towards the ulnar side.
Grades 2, 1 and 0 (see Fig. 3.53, p. 57): With the supported forearm mid-way between supination and pronation, so that the ulnar surface of the forearm is resting on the support, the wrist is flexed.

Comments: When testing wrist flexion the fingers should be relaxed. Finger flexion during the test indicates substitution of flexor digitorum profundus or flexor digitorum superficialis.

FLEXOR CARPI RADIALIS

Attachments: Extends from the medial epicondyle of the humerus to the anterior surface of the base of the second and third metacarpals.
Nerve supply: Median C6, 7.

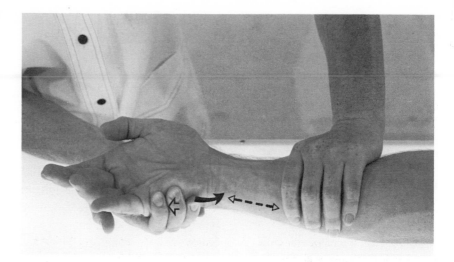

Fig. 3.54

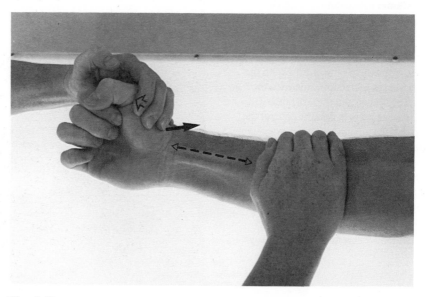

Fig. 3.55

Surface markings: It is the most lateral wrist flexor tendon and can be palpated as it crosses the anterior surface of the wrist joint.

Actions
Rostral end fixed: It flexes and abducts the wrist. It may also assist pronation of the forearm and flexion of the elbow.
Caudal end fixed: It assists in stabilizing the wrist joint during weight bearing through the tips of the digits, as in finger-tip push-ups.

Maximum extensibility: With the elbow extended, the wrist joint is radially deviated and extended.

TESTING POSITION
Grades 5, 4 and 3 (Fig. 3.55): With the forearm supported in supination, the wrist is flexed in a radial direction. Resistance is applied to the thenar eminence in the direction of wrist extension and ulna deviation.
Grades 2, 1 and 0 (see Fig. 3.53, p. 57): With the forearm supported and the thumb uppermost, the wrist is flexed towards the radial side.

Comments: When testing for a Grade 1, care must be taken not to mistake the radial pulse for a flicker of movement.

EXTENSOR CARPI RADIALIS LONGUS

Attachments: Extends from the lateral epicondyle of the humerus to the posterior surface of the base of the second metacarpal bone.
Nerve supply: Radial C6, 7.

Surface markings: It can be palpated in the anatomical snuff-box as it crosses the posterior surface of the wrist joint to insert into the base of the second metacarpal. In can also be palpated at the elbow, above the lateral epicondyle.

EXTENSOR CARPI RADIALIS BREVIS

Attachments: Extends from the lateral epicondyle of the humerus to the posterior surface of the base of the third metacarpal.
Nerve supply: Posterior interosseus C7, 8.

Surface markings: This muscle can be palpated in the anatomical snuff-box on the posterolateral surface of the wrist at the base of the third metacarpal. It can also be palpated at the elbow just distal to the lateral epicondyle on the posterior surface of the forearm.

For both radial wrist extensors

Actions
Rostral end fixed: They extend and abduct the wrist, and may assist also in flexion of the elbow.
Caudal end fixed: They assist in stabilizing the wrist joint during weight bearing through the tips of the digits, as in finger-tip push-ups.

Maximum extensibility: With the elbow extended, the wrist is flexed and ulna deviated.

TESTING POSITION
Grades 5, 4 and 3 (Fig. 3.56): With the elbow flexed to 30° and the supported forearm pronated, the wrist is extended in a radial direction. Resistance is applied on the posterior surface of the second and third metacarpal bones in the direction of flexion and ulna deviation.

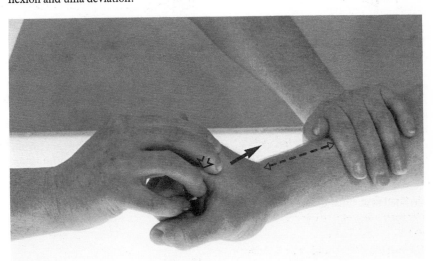

Fig. 3.56

Grades 2, 1 and 0 (not illustrated): With the elbow flexed to about 30° and the forearm supported with the thumb uppermost, the wrist is extended in a radial direction.

Comments: When weakness is present, extensor pollicis longus and brevis may be used to produce this movement. Extension of the thumb will suggest that this substitution is occurring.

Muscles tested at the wrist joint

EXTENSOR CARPI ULNARIS

Attachments: Extends from the lateral epicondyle of the humerus and the posterior surface of the ulna to the base of the fifth metacarpal.
Nerve supply: Posterior interosseous C7, 8.

Surface markings: The tendon can be palpated on the posterior surface of the wrist, just proximal to the base of the fifth metacarpal.

Actions
Rostral end fixed: It extends and adducts the wrist.
Caudal end fixed: It assists in stabilizing the wrist joint during weight bearing through the tips of the digits, as in finger-tip push-ups.

Maximum extensibility: With the elbow extended, the wrist joint is flexed and radially deviated.

TESTING POSITION
Grades 5, 4 and 3 (Fig. 3.57): With the pronated forearm supported, the wrist is extended in an ulnar direction. Resistance is applied to the posterior surface of fourth and fifth metacarpal bones in the direction of flexion and radial deviation.
Grades 2, 1 and 0 (not illustrated): With the supported forearm midway between pronation and supination, the wrist is extended in an ulnar direction.

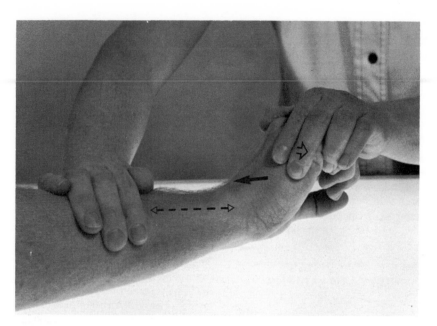

Fig. 3.57

Comments: The metacarpophalangeal joints tend to flex during this movement. Where they begin to extend, substitution of extensor digitorum longus should be suspected. Excessive finger extension during testing is suggestive of weakness.

Muscles tested at the joints of the fingers (Figs 3.58 & 3.59)

Fig. 3.58

Fig. 3.59

FLEXION OF THE METACARPOPHALANGEAL JOINTS
Lumbricals
Flexor digiti minimi (FDM)

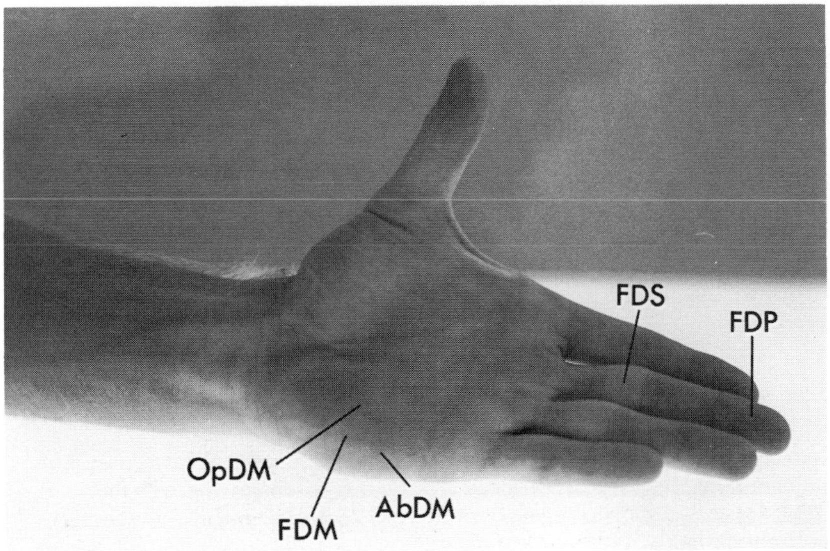

FLEXION OF THE PROXIMAL INTERPHALANGEAL JOINTS
Flexor digitorum superficialis (FDS)

FLEXION OF THE DISTAL INTERPHALANGEAL JOINTS
Flexor digitorum profundus (FDP)

ABDUCTION OF THE METACARPOPHALANGEAL JOINTS
Abductor digiti minimi (AbDM)
Dorsal interossei

ADDUCTION OF THE METACARPOPHALANGEAL JOINTS
Palmar interossei

EXTENSION OF THE METACARPOPHALANGEAL JOINTS
Extensor digitorum (ED)
Extensor indicis (EI)
Extensor digiti minimi (EDM)

EXTENSION OF THE PROXIMAL INTERPHALANGEAL JOINTS
Extensor digitorum
Lumbricals

OPPOSITION OF THE FIFTH DIGIT
Opponens digiti minimi (OpDM)

Note: Because the muscles which move fingers and thumb are comparatively small, the tester often has a mechanical advantage over the subject. It is therefore important, in determining strength, to compare the muscles being tested with those of the other hand, or with those of a person of similar age, sex, build, and occupation.

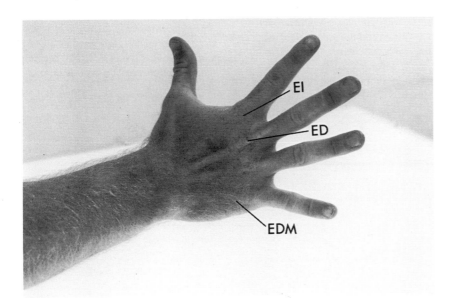

Upper extremity

Muscles tested at the joints of the fingers

LUMBRICALS

Attachments: The two lateral components originate from the radial side of the tendons of flexor digitorum profundus (to the first two fingers) and the two medial components from the adjacent sides of the second and third, and the third and fourth tendons of flexor digitorum profundus (to the third and fourth fingers). Each lumbrical inserts into the radial side of the extensor expansion of the proximal phalanx of the respective digits.

Nerve supply: The lateral two, median C8, T1. The medial two, ulnar C8, T1.

Surface markings: The first lumbrical is the only one easily palpated. It can be found on the lateral side of the first digit just below the metacarpophalangeal joint.

Actions

Rostral end fixed: These muscles flex the metacarpophalangeal joints and extend the interphalangeal joints of the second to fifth digits.

Caudal end fixed: These muscles assist in stabilizing the metacarpophalangeal and interphalangeal joints during weight bearing through the tips of the digits, as in finger-tip push-ups.

Maximum extensibility: With metacarpophalangeal joints of the second to fifth digits extended, the interphalangeal joints are flexed.

TESTING POSITION

Grades 5, 4 and 3 (Fig. 3.60): In sitting or supine lying, with the forearm in a vertical position supported through the elbow and the wrist fixed, the

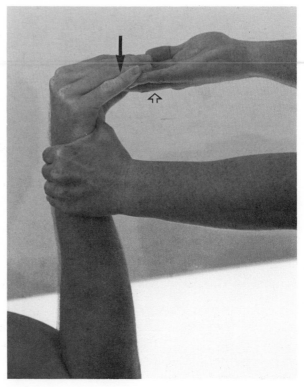

Fig. 3.60

metacarpophalangeal joints are flexed and the interphalangeal joints extended. Resistance is applied to the anterior surface of the first to fifth digits in the direction of extension of the metacarpophalangeal joints. Alternatively, a piece of paper can be held between the fingers and opposed thumb with the metacarpophalangeal joints flexed and the interphalangeal joints extended. The tester then tries to remove it from the subject's grip. It is important that the wrist is fixed during this test.

Grades 2, 1 and 0 (Not illustrated): In sitting or supine lying, with the supported forearm mid-way between pronation and supination and the wrist in a neutral position, the metacarpophalangeal joints are flexed and the interphalangeal joints extended.

Comments: If the interphalangeal joints are not maintained in extension throughout the test, this is indicative of weakness of the lumbricals. The position described for Grades 5, 4 and 3 is not against gravity.

FLEXOR DIGITI MINIMI

Attachments: Extends from the hook of the hamate and flexor retinaculum to the ulnar side of the base of the proximal phalanx of the little finger.
Nerve supply: Ulnar C8, T1.

Surface markings: It can be palpated near its distal attachment.

Actions
Rostral end fixed: It flexes the proximal phalanx of the little finger. It also assists in opposition of the little finger.
Caudal end fixed: It assists in stabilizing the fifth metacarpophalangeal joint during weight bearing through the tips of the digits, as in finger-tip push-ups.

Maximum extensibility: With the wrist extended and radially deviated, the metacarpophalangeal joint of the little finger is extended.

TESTING POSITION
Grades 5, 4 and 3 (Fig. 3.61): In sitting or supine lying, with the supported forearm supinated, and the hand and wrist fixed, the proximal phalanx of the little finger is flexed. Resistance is applied to the anterior surface of the proximal

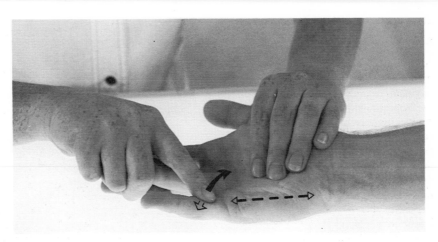

Fig. 3.61

phalanx in the direction of extension.
Grades 2, 1 and 0 (Not illustrated): In sitting or supine lying, with the forearm mid-way between pronation and supination, and the wrist and metacarpals fixed in a neutral position, the little finger is flexed.

Comments: Without adequate stabilization of the wrist and metacarpals it will be very difficult to elicit a Grade 2 contraction in a weak muscle.

Upper extremity

Muscles tested at the joints of the fingers

FLEXOR DIGITORUM SUPERFICIALIS

Attachments: The humero-ulnar head originates from the medial epicondyle of the humerus, and the medial side of the coronoid process of the ulnar, and the radial head from the oblique line of the radius. Both heads insert by four tendons into both sides of the middle phalanx of the second to fifth digits.
Nerve supply: Median C7, 8, T1.

Surface markings: The muscle belly may be palpated on the anterior aspect of the forearm, and the tendon to the ring finger may be identified on the anterior surface of the wrist joint between the tendons of palmaris longus and flexor carpi ulnaris.

Actions
Rostral end fixed: It flexes the proximal interphalangeal joint of the second to fifth digits. It also assists in flexion of the metacarpophalangeal joints of the same digits and flexion of the wrist joint.
Caudal end fixed: It assists in stabilizing the metacarpophalangeal joints during weight bearing through the tips of the digits, as in finger-tip push-ups.

Maximum extensibility: With the elbow extended the wrist and second to fifth digits are extended.

TESTING POSITION
Grades 5, 4 and 3 (Fig. 3.62): In sitting or supine lying, with the supported forearm supinated and the metacarpal bone of the digit being tested fixed, the proximal interphalangeal joint is flexed while the distal one remains extended. Resistance is applied to the anterior surface of the

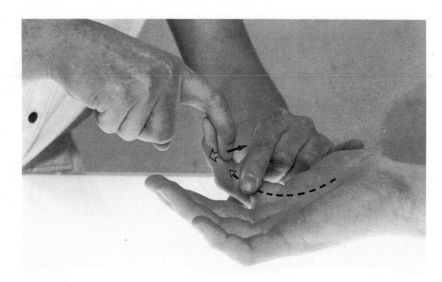

Fig. 3.62

middle phalanx in the direction of extension. Each digit is tested separately.
Grades 2, 1 and 0 (Same position as for Fig. 3.64, p. 65): In sitting or supine lying, with the supported forearm midway between pronation and supination and the wrist in a neutral position, the metacarpal bone of the digit being tested is fixed and the proximal interphalangeal joint is flexed with the distal one extended. Each digit is tested separately.

Comments: The wrist must be maintained in a neutral position throughout testing to ensure that no undue tension on the long extensors limits the movement of the distal phalanx. Also, it may be necessary to stabilize the wrist when some degree of weakness is present.

FLEXOR DIGITORUM PROFUNDUS

Attachments: Originates from the anterior surface of the upper third of the ulna. It inserts by four tendons onto the anterior surface of the base of the distal phalanx of the second to fifth digits.

Nerve supply: Anterior interosseous C8, T1 to the lateral half of the muscle (index and middle fingers). Ulnar C8, T1 to the medial half of the muscle (ring and little fingers).

Surface markings: The tendons may be palpated on the palmar surface of the middle phalanx.

Actions

Rostral end fixed: It flexes the distal phalanx of digits two to five. Also, it assists in flexion of the metacarpophalangeal joints of each of these digits and flexion of the wrist.

Caudal end fixed: It assists in stabilizing the metacarpophalangeal joints during weight bearing through the tips of the digits, as in finger-tip push-ups.

Maximum extensibility: With the elbow and wrist extended, the metacarpophalangeal and interphalangeal joints are extended such that weight can be taken on the palmar surface of the hand.

TESTING POSITION

Grades 5, 4 and 3 (Fig. 3.63): In sitting or supine lying, with the supported forearm supinated and the proximal and middle phalanges of the digit being tested fixed, the distal phalanx is flexed. Resistance is applied to the anterior surface of the distal phalanx in the direction of extension. Each digit is tested separately.

Grades 2, 1 and 0 (Fig. 3.64): In sitting or supine lying, with the supported forearm mid-way between pronation and supination, and the proximal and middle

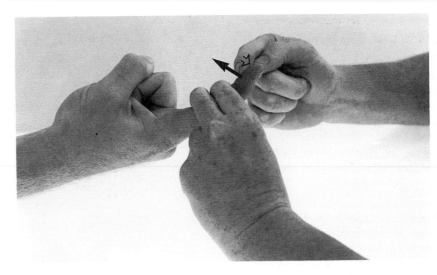

Fig. 3.63

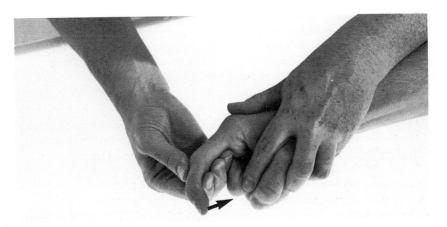

Fig. 3.64

phalanges of the digit being tested fixed, the distal phalanx is flexed. Each digit is tested separately.

Comments: The wrist must be maintained in a neutral position throughout testing to ensure that no undue tension on the long extensors limits the movement of the distal phalanx. When testing for Grades 5, 4 and 3, the wrist may require stabilization, if some degree of muscle weakness is present.

Muscles tested at the joints of the fingers

ABDUCTOR DIGITI MINIMI

Attachments: Extends from the pisiform bone to the ulna surface of the base of the proximal phalanx of the little finger and the ulna border of the extensor expansion.
Nerve supply: Ulnar C8,T1.

Surface markings: The muscle can be palpated on the ulna border of the hand.

Actions
Rostral end fixed: It abducts the little finger. It may assist in opposition and flexion of the little finger also.
Caudal end fixed: It stabilizes the fifth metacarpophalangeal joint during weight bearing through the tips of the digits, as in finger-tip push-ups.

Maximum extensibility: With the wrist radially deviated and extended, the little finger is adducted.

TESTING POSITION
Grades 5, 4 and 3 (Fig. 3.65): In sitting or supine lying, with the supported forearm supinated and the hand and wrist fixed, the little finger is abducted. Resistance is applied to the ulnar border of the little finger in the direction of adduction.

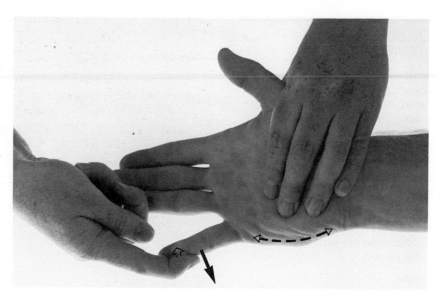

Fig. 3.65

Grades 2, 1 and 0 (Not illustrated): In sitting or supine lying, with the supported forearm supinated and the hand and wrist fixed, the little finger is abducted.

Comments: The test position for Grades 5, 4 and 3 is not against gravity as this is not practical.

DORSAL INTEROSSEI

Attachments: Each of the four muscles extends by two heads from the shafts of the adjacent metacarpal bones in each interspace to the extensor expansion and base of the proximal phalanx on the radial side, in the case of the index and middle fingers, and on the ulnar side for the middle and ring fingers.
Nerve supply: Ulnar C8, T1.

Surface markings: These muscles can be palpated deep in the web spaces between the fingers. The first dorsal interosseus is the most accessible on the lateral aspect of the first metacarpophalangeal joint.

Actions
Rostral end fixed: Abduction of index, middle and ring fingers away from the axis of the middle finger. These muscles also assist in flexion of the metacarpophalangeal joints and, through the extensor expansion, extend the interphalangeal joints of the same fingers.
Caudal end fixed: These muscles assist in stabilization of the metacarpophalangeal and interphalangeal joints during weight bearing through the tips of the digits, as in finger-tip push-ups.

Maximum extensibility: With the metacarpophalangeal and the interphalangeal joints in extension, each finger is adducted.

TESTING POSITION
Grades 5, 4 and 3 (Fig. 3.66): In sitting or supine lying, with the supported forearm

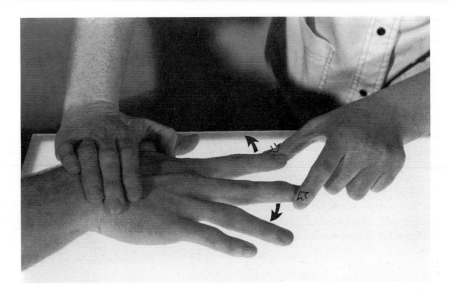

Fig. 3.66

pronated and the palm of the hand flat on the supporting surface, the index, middle, and ring fingers are abducted. Resistance is applied to the lateral surface of the distal phalanx of the fingers being tested, in the direction of adduction. Each finger should be tested separately.
Grades 2, 1 and 0 (Not illustrated): In sitting or supine lying, with the supported forearm pronated and the palm of the hand flat on the supporting surface, the index, middle and ring fingers are abducted. Each finger should be tested separately. Figure 3.66 shows the testing position.

Comments: The position described for Grades 5, 4 and 3 is not against gravity.

Upper extremity

Muscles tested at the joints of the fingers

PALMAR INTEROSSEI

Attachments: The first originates from the ulnar side of the base of the first metacarpal, the second from the ulnar side of the shaft of the second metacarpal and the third and fourth from the radial side of the shafts of the third and fourth metacarpal bones, respectively. The first and second insert in the ulnar side of the extensor expanion and base of the proximal phalanx of the same digits and the third and fourth into the radial side of the extensor expansion and base of the proximal phalanx of the same digits.
Nerve supply: Ulnar C8, T1.

Surface markings: Palpation deep in the web spaces is possible, but the contraction of each palmar and dorsal interosseus can be difficult to distinguish.

Actions
Rostral end fixed: These muscles adduct the digits towards the axis of the middle finger. They assist in flexion of the metacarpophalangeal joint and, through the extensor expansion, extend the interphalangeal joints of the index, ring and little fingers.
Caudal end fixed: These muscles help to stabilize the interphalangeal joints during weight bearing through the tips of the digits, as in finger-tip push-ups.

Maximum extensibility: With the metacarpophalangeal and interphalangeal joints extended, each finger is abducted.

TESTING POSITION
Grades 5, 4 and 3 (Fig. 3.67): In sitting or supine lying, with the supported forearm supinated and the dorsum of the hand

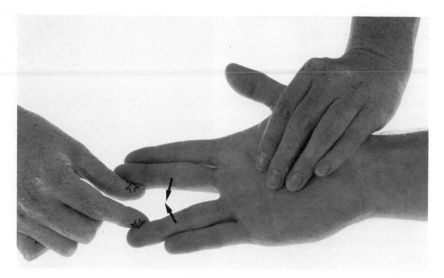

Fig. 3.67

flat on the supporting surface, the index, ring, and little fingers are adducted towards the middle finger. Resistance is applied to the lateral surface of the distal phalanx of the fingers being tested, in the direction of abduction. Each finger should be tested separately.

Alternatively, the tester can attempt to remove a piece of paper gripped between each of the extended fingers in turn.
Grades 2, 1 and 0 (Not illustrated): In sitting or supine lying, with the supported forearm supinated and the dorsum of the hand flat on the supporting surface, the index, ring and middle fingers are adducted toward the middle finger. Each finger should be tested separately.

Comments: The position described for Grades 5, 4 and 3 is not against gravity. This muscle can also be tested with the forearm in pronation with the palm resting on the supporting surface.

EXTENSOR DIGITORUM

Attachments: Originates from the lateral epicondyle of the humerus. It divides into four tendons and inserts via a medial band into the base of the middle phalanx of each of the second to fifth digits and via two lateral bands into the base of the distal phalanx of the same digits.

Nerve supply: Posterior interosseous C7, 8.

Surface markings: The four tendons can be palpated on the posterior surface of the metacarpophalangeal joints of the second to fifth digits.

Actions

Rostral end fixed: It extends the metacarpophalangeal joints and assists in extension of the wrist. Together with the lumbricals and interossei it extends the interphalangeal joints of the second to fifth digits. With the wrist extended, this muscle can assist in abduction of the index, ring, and little fingers.

Caudal end fixed: It assists in stabilizing the metacarpophalangeal joints during weight bearing through the tips of the digits, as in finger-tip push-ups.

Maximum extensibility: With the elbow joint extended and the forearm pronated, the wrist and fingers are flexed.

TESTING POSITION

Grades 5, 4 and 3 (Fig. 3.68): In sitting or supine lying, with the supported forearm pronated and the wrist in extension, the metacarpophalangeal joints of the second to fifth digits are extended, with the interphalangeal joints of the same fingers in flexion. Resistance is applied to the posterior surface of the proximal phalanges of the second to fifth digits in the direction of flexion.

Grades 2, 1 and 0 (Fig. 3.69): In sitting or supine lying, with the forearm supported mid-way between pronation and supination the metacarpophalangeal joints of the second to fifth digits are extended with the interphalangeal joints of the same fingers in flexion.

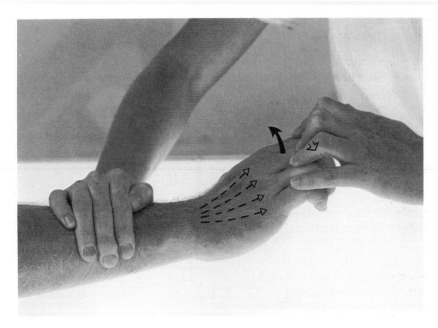

Fig. 3.68

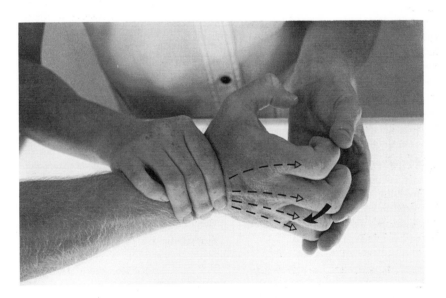

Fig. 3.69

Comments: It is essential that the wrist be well stabilized when testing metacarpophalangeal joint extension, particularly for Grades 2, 1 and 0. Otherwise, activity of the wrist extensors may be substituted for that of extensor digitorum.

Muscles tested at the joints of the fingers

EXTENSOR INDICIS

Attachments: Extends from the posterior surface of the lower third of the ulnar and interosseous membrane to the posterior surface of the first phalanx of the index finger and the extensor expansion.
Nerve supply: Posterior interosseous C7, 8.

Surface markings: The tendon can be palpated over the posterior surface of the first metacarpophalangeal joint. It is medial to the tendon of extensor digitorum for the index finger.

Actions
Rostral end fixed: It extends the metacarpophalangeal joint of the index finger. With the lumbricals and interossei working through the extensor expansion, it extends the other joints of the same finger. It assists in adduction of the index finger.
Caudal end fixed: It assists in stabilizing the metacarpophalangeal and interphalangeal joints when weight bearing through the tips of the digits, as in finger-tip push-ups.

Maximum extensibility: With the forearm pronated and the wrist flexed, the first metacarpophalangeal and interphalangeal joints are flexed.

TESTING POSITION
Grades 5, 4 and 3 (Fig. 3.70): In sitting or supine lying, with the supported forearm pronated and the wrist in about 10° of extension, the metacarpophalangeal joint of the index finger is extended.

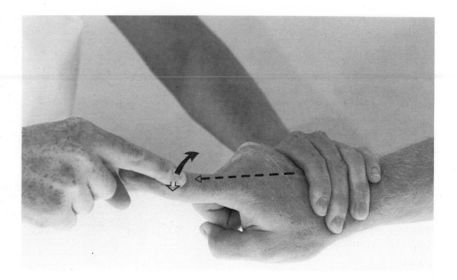

Fig. 3.70

Resistance is applied to the posterior surface of the proximal phalanx of the index finger in the direction of flexion.
Grades 2, 1 and 0 (Not illustrated): In sitting or supine lying, with the supported forearm mid-way between pronation and supination, the metacarpophalangeal joint of the index finger is extended.

Comments: The lateral three digits should be maintained in flexion while testing to inhibit the action of extensor digitorum. As the interossei can assist the action of this muscle, flexion of the interphalangeal joints will further isolate the action of the muscle when weakness is present.

EXTENSOR DIGITI MINIMI

Attachments: Extends from the lateral epicondyle of the humerus to the extensor expansion of the little finger.
Nerve supply: Posterior interosseous C7, 8.

Surface markings: The tendon can be palpated on the dorsal surface of the hand. It is the extensor tendon closest to the ulna border of the hand.

Actions

Rostral end fixed: It extends the metacarpophalangeal joint of the little finger. Along with the lumbricals and interossei acting through the extensor expansion, it also extends the other joints of the same finger. It assists in abduction of the little finger.
Caudal end fixed: It stabilizes the fifth metacarpophalangeal joint during weight bearing through the tips of the digits, as in finger-tip push-ups.

Maximum extensibility: With the wrist flexed, the metacarpophalangeal joint of the little finger is flexed.

TESTING POSITION

Grades 5, 4 and 3 (Fig. 3.71): In sitting or supine lying, with the supported forearm pronated and the wrist in a neutral position, the metacarpophalangeal joint of the little finger is extended, with the interphalangeal joints in extension.

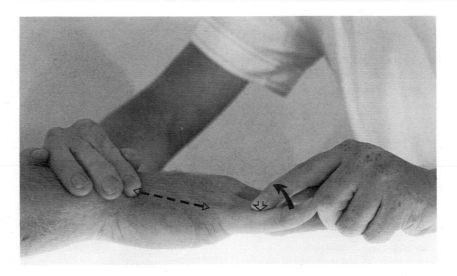

Fig. 3.71

Resistance is applied to the posterior surface of the proximal phalanx of the little finger in the direction of flexion.
Grades 2, 1 and 0 (Not illustrated): In sitting or supine lying, with the supported forearm mid-way between pronation and supination, the metacarpophalangeal joint of the little finger is extended with the interphalangeal joints in flexion.

Comments: Where there is weakness of the extensor digiti minimi, it may be difficult to differentiate it from extensor digitorum.

Upper extremity

Muscles tested at the joints of the fingers

OPPONENS DIGITI MINIMI

Attachments: Extends from the hook of the hamate and the flexor retinaculum to the ulnar side of the shaft of the fifth metacarpal.
Nerve supply: Ulnar C8, T1.

Surface markings: It is difficult to palpate as it is deep to both the flexor and abductor digiti minimi.

Actions
Rostral end fixed: It opposes the fifth metacarpal and draws the ulnar border of the hand towards the centre of the palm.
Caudal end fixed: It stabilizes the metacarpophalangeal joint during weight bearing through the tips of the digits, as in finger-tip push-ups.

Maximum extensibility: With the wrist extended the metacarpophalangeal joint of the little finger is abducted and extended.

TESTING POSITION
Grades 5, 4 and 3 (Fig. 3.72): In sitting or supine lying, with the supported forearm supinated and the hand and wrist fixed, the fifth metacarpal is opposed towards the first. Resistance is applied to the anterior surface of the distal end of the fifth metacarpal in the direction of the supporting surface.
Grades 2, 1 and 0 (Not illustrated): In sitting or supine lying, with the forearm supported at the elbow so that the forearm is vertical, the fifth metacarpal is opposed towards the first. The forearm, wrist and hand must be fixed.

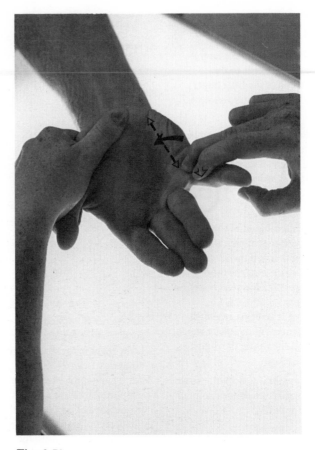

Fig. 3.72

Comments: Adequate fixation of the forearm, wrist, and hand when testing for Grades 2, 1 and 0 is necessary whenever weakness is present.

Muscles tested at the joints of the thumb (Figs 3.73 & 3.74)

FLEXION OF THE
METACARPOPHALANGEAL JOINT
Flexor pollicis brevis (FPB)
Flexor pollicis longus

FLEXION OF THE
INTERPHALANGEAL JOINT
Flexor pollicis longus (FPL)

EXTENSION OF THE
METACARPOPHALANGEAL JOINT
Extensor pollicis brevis (EPB)
Extensor pollicis longus

EXTENSION OF THE
INTERPHALANGEAL JOINT
Extensor pollicis longus (EPL)

ABDUCTION OF THE
METACARPOPHALANGEAL JOINT
Abductor pollicis brevis (AbPB)
Abductor pollicis longus

ABDUCTION OF THE
INTERPHALANGEAL JOINT
Abductor pollicis longus (AbPL)

ADDUCTION OF THE
METACARPOPHALANGEAL AND
INTERPHALANGEAL JOINTS
Adductor pollicis (AdP)

OPPOSITION OF THE THUMB
Opponens pollicis (OpP)

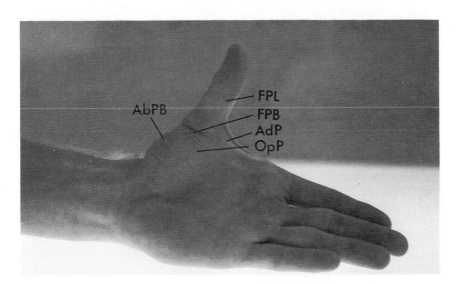

Fig. 3.73

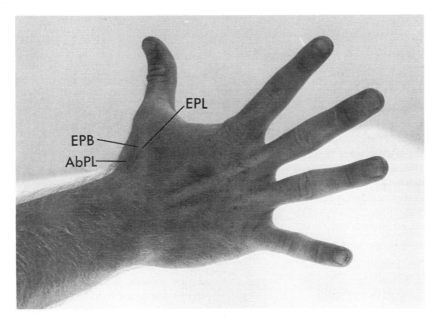

Fig. 3.74

Muscles tested at the joints of the thumb

FLEXOR POLLICIS BREVIS

Attachments: Extends from the flexor retinaculum and the ulnar border of the first metacarpal to the base of the first phalanx of the thumb.
Nerve supply: Median C6, 7, 8, T1 and occasionally by the ulnar C8, T1.

Surface markings: It can be palpated on the palmar surface of the first metacarpophalangeal joint.

Actions
Rostral end fixed: It flexes the proximal phalanx of the thumb.
Caudal end fixed: It assists in stabilizing the metacarpophalangeal joint during weight bearing through the tips of the digits, as in finger-tip push-ups.

Maximum extensibility: With the wrist extended, the thumb is extended.

TESTING POSITION
Grades 5, 4 and 3 (Fig. 3.75): With supported forearm supinated, the first metacarpal and wrist fixed, the proximal phalanx of the thumb is flexed across the palm of the hand. Resistance is applied to the palmar surface of the proximal phalanx of the thumb in the direction of extension.
Grades 2, 1 and 0 (Not illustrated): The position and test is the same as for Grades 5, 4 and 3 except the interphalangeal joint should be extended to inhibit the action of flexor pollicis longus, and no resistance is applied.

Comments: Because of the difficulty involved in assuming a gravity resisted position, (internal rotation of the shoulder and pronation of the forearm), the muscle is tested in the gravity eliminated position for all grades.

FLEXOR POLLICIS LONGUS

Attachments: Extends from the anterior surface of the middle third of the radius to the anterior surface of the base of the distal phalanx of the thumb.
Nerve supply: Anterior interosseous C8, T1.

Surface markings: Its tendon can be palpated as it crosses the palmar surface of the interphalangeal joint of the thumb.

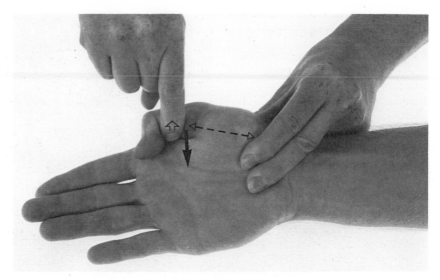

Fig. 3.75

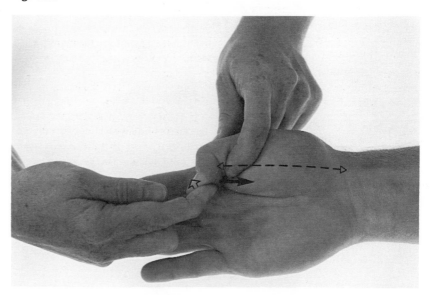

Fig. 3.76

Actions
Rostral end fixed: It flexes the distal phalanx of the thumb. It assists in flexion of the other joints of the thumb and in flexion of the wrist.
Caudal end fixed: It assists in stabilizing the thumb and wrist during weight bearing through the tips of the digits, as in finger-tip push-ups.

Maximum extensibility: With the forearm supinated and the wrist extended, the thumb is extended.

TESTING POSITION
Grades 5, 4 and 3 (Fig. 3.76): In sitting or supine lying, with the supported forearm supinated and the first metacarpal and proximal phalanx of the thumb fixed, the distal phalanx of the thumb is flexed. Resistance is applied to the palmar surface of the distal phalanx of the thumb in the direction of extension.
Grades 2, 1 and 0 (Not illustrated): The position and test is the same as for Grades 5, 4 and 3, but no resistance is applied.

Comments: Because of the difficulty involved in assuming a gravity resisted position (shoulder flexion and internal rotation with forearm pronation), the muscle is tested in the gravity eliminated position for all grades.

EXTENSOR POLLICIS BREVIS

Attachments: Extends from the distal end of the posterior surface of the radius and the interosseous membrane to the posterior surface of the base of the proximal phalanx of the thumb.

Nerve supply: Posterior interosseous C7, 8.

Surface markings: The tendon can be palpated at the lateral border of the anatomical snuff box.

Actions

Rostral end fixed: It extends the metacarpophalangeal joint of the thumb. It assists also in extension and radial deviation of the wrist and carpometacarpal joints.

Caudal end fixed: It assists in stabilizing the wrist and thumb during weight bearing through the tips of the digits, as in finger-tip push-ups.

Maximum extensibility: With the forearm pronated and the wrist flexed, the thumb is flexed across the palm of the hand.

TESTING POSITION

Grades 5, 4 and 3 (Fig. 3.77): In sitting or supine lying, with the supported forearm mid-way between pronation and supination, the metacarpophalangeal joint of the thumb is extended away from the radial side of the index finger.

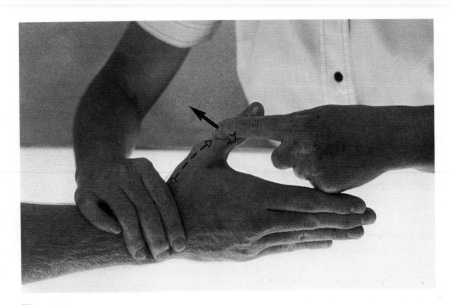

Fig. 3.77

Resistance is applied to the posterior surface of the proximal phalanx of the thumb in the direction of flexion.

Grades 2, 1 and 0 (Not illustrated): In sitting or supine lying, with the supported forearm pronated, the metacarpophalangeal joint of the thumb is extended away from the radial side of the index finger.

Comments: The proximal phalanx and the rest of the hand and wrist must be well fixed during the test procedure, particularly if this muscle is weak.

Muscles tested at the joints of the thumb

EXTENSOR POLLICIS LONGUS

Attachments: It extends from the middle third of the posterior surface of the ulna and the interosseous membrane to the posterior surface of the base of the proximal phalanx of the thumb.

Nerve supply: Posterior interosseous C7, 8.

Surface markings: The tendon can be palpated at the medial border of the anatomical snuff box.

Actions

Rostral end fixed: It extends all joints of the thumb and assists in extension of the wrist.

Caudal end fixed: It assists in stabilizing the wrist and thumb during weight bearing through the tips of the digits, as in finger-tip push-ups.

Maximum extensibility: With the forearm pronated and the wrist flexed, the thumb is flexed across the palm of the hand.

TESTING POSITION

Grades 5, 4 and 3 (Fig. 3.78): In sitting or supine lying, with the supported forearm mid-way between pronation and supination, the interphalangeal joint of the thumb is extended away from the radial side of the index finger. Resistance is applied to the posterior surface of the distal phalanx of the thumb in the direction of flexion.

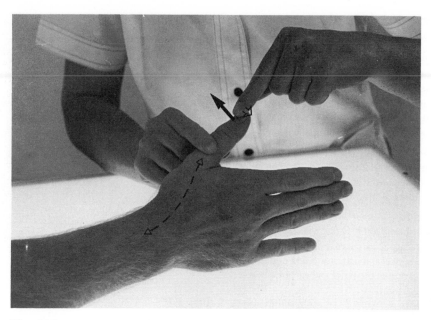

Fig. 3.78

Grades 2, 1 and 0 (Not illustrated): In sitting or supine lying, with the supported forearm pronated, the interphalangeal joint of the thumb is extended away from the radial side of the index finger.

Comments: The proximal phalanx and the rest of the hand and wrist must be well fixed during the test procedure, especially if this muscle is weak.

ABDUCTOR POLLICIS BREVIS

Attachments: Extends from the ridge of the scaphoid and the flexor retinaculum to the lateral side of the base of the proximal phalanx of the thumb.
Nerve supply: Median C8, T1

Surface markings: The muscle can be palpated on the lateral part of the thenar eminence.

Actions
Rostral end fixed: It abducts the thumb.
Caudal end fixed: It assists in stabilizing the thumb and wrist during weight bearing on the tips of the digits, as in finger-tip push-ups.

Maximum extensibility: With the wrist extended, the thumb is adducted.

TESTING POSITION
Grades 5, 4 and 3 (Fig. 3.79): In sitting or supine lying, with the supported forearm supinated, the thumb is abducted anteriorly away from the palm of the hand. Resistance is applied to the lateral side of the proximal phalanx of the thumb in the direction of adduction.
Grades 2, 1 and 0 (Not illustrated): In sitting or supine lying, with the forearm supported and in a mid-position, the thumb is abducted away from the palm of the hand.

Comments: If there is marked weakness of this muscle it is important to stabilise both the wrist and metacarpals to prevent unwanted movement.

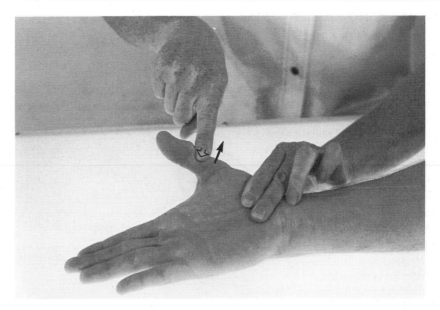

Fig. 3.79

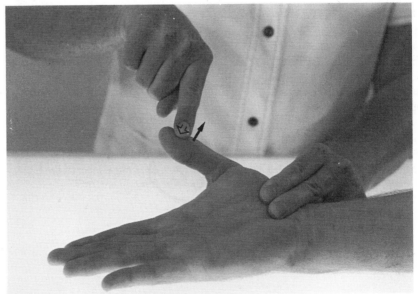

Fig. 3.80

ABDUCTOR POLLICIS LONGUS

Attachments: Extends from the interosseous membrane and the posterior surface of the middle third of the radius and ulna to the lateral side of the base of the first metacarpal.
Nerve supply: Posterior interosseous C7, 8.

Surface markings: The tendon can be palpated at the base of the first metacarpal lateral to the tendon of extensor pollicis brevis.

Actions
Rostral end fixed: It abducts the thumb and assists in extension of the thumb.

Caudal end fixed: It assists in stabilizing the thumb and wrist during weight bearing on the tips of the digits, as in finger-tip push-ups.

Maximum extensibility: With the forearm pronated and the wrist flexed and ulna deviated, the thumb is adducted.

TESTING POSITION
Grades 5, 4 and 3 (Fig. 3.80): In sitting or supine lying, with the supported forearm supinated, the thumb is abducted in an anterior direction away from the palm of the hand. Resistance is applied to the lateral surface of the distal phalanx of the thumb in the direction of adduction.
Grades 2, 1 and 0 (Not illustrated): In sitting or supine lying, with the supported forearm mid-way between pronation and supination, the thumb is abducted away from the palm of the hand.

Comments: If there is marked weakness of this muscle it is important to stabilize both the wrist and metacarpals to prevent unwanted movement.

Muscles tested at the joints of the thumb

ADDUCTOR POLLICIS

Attachments: The transverse head originates from the palmar surface of the shaft of the third metacarpal, and the oblique head from the base of the second and third metacarpal, the trapezoid and capitate. Both heads insert into the medial side of the proximal phalanx of the thumb and the extensor expansion. *Nerve supply*: Ulnar C8, T1.

Surface markings: It can be palpated in the web space between the thumb and index fingers.

Actions
Rostral end fixed: It adducts the thumb and assists in opposition.
Caudal end fixed: It assists in stabilizing the thumb during weight bearing through the tips of the digits, as in finger-tip push-ups.

Maximum extensibility: With the metacarpophalangeal joints of the fingers fully extended, the thumb is abducted in an anterior direction away from the palm.

TESTING POSITION
Grades 5, 4 and 3 (Fig. 3.81): In sitting or supine lying, with the supported forearm pronated and the fixed wrist and fingers extended, the thumb is adducted towards the palm of the hand. Resistance is

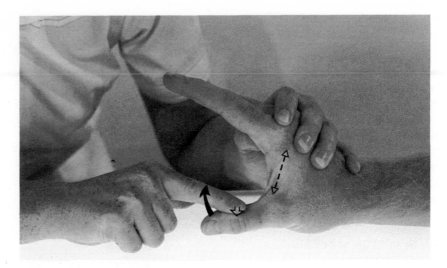

Fig. 3.81

applied to the lateral surface of the interphalangeal joint of the thumb in the direction of abduction.
Grades 2, 1 and 0 (Not illustrated): In sitting or supine lying, with the supported forearm mid-way between pronation and supination, the thumb is adducted toward the palm.

Comments: Where weakness occurs, stabilization of both the wrist and metacarpals is most important to limit unwanted movement.

OPPONENS POLLICIS

Attachments: Extends from the tubercle of the trapezium and the flexor retinaculum to the anterior surface of the shaft of the first metacarpal.
Nerve supply: Median C8, T1.

Surface markings: It lies under the abductor pollicis brevis and is difficult to palpate.

Actions
Rostral end fixed: It opposes the thumb to the other fingers.
Caudal end fixed: It assists in stabilizing the thumb during weight bearing through the tips of the digits, as in finger-tip push-ups.

Maximum extensibility: With the forearm supinated, the thumb is extended and adducted.

TESTING POSITION
Grades 5, 4 and 3 (Fig. 3.82): In sitting or supine lying, with the supported forearm supinated, the thumb is opposed towards the little finger. Resistance is applied to the anterior surface of the proximal phalanx in the opposite direction.
Grades 2, 1 and 0 (Not illustrated): In sitting or supine lying, with the forearm vertical and supported on the elbow, the thumb is opposed towards the little finger.

Comments: As the thumb can be moved across the palm of the hand by flexor

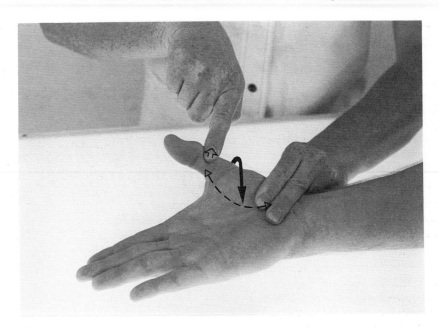

Fig. 3.82

pollicis longus and brevis, it is important to determine whether this action is being performed by these muscles or opponens pollicis. If the first metacarpal undergoes any medial rotation (i.e. the thumb nail shows a palmar view) as the thumb is moved towards the little finger, then opponens pollicis is functioning. In the patient with any weakness of this muscle, stabilization of both the wrist and the metacarpals is essential if the action is to be isolated.

4. Evaluating the muscles of the lower extremity

Muscles tested at the hip joint (Figs 4.1 & 4.2)

EXTENSION OF THE HIP JOINT
Gluteus maximus (GMax)
   Biceps femoris
   Semimembranosus
   Semitendinosus

FLEXION OF THE HIP JOINT
Psoas major   } **Iliopsoas (IP)**
Iliacus
   Rectus femoris
Tensor fasciae latae (TFL)
Sartorius (S)

ADDUCTION OF THE HIP JOINT
Adductor magnus (AM)
Adductor longus (AL)
Adductor brevis (AB)
Gracilis (G)
Pectineus

ABDUCTION OF THE HIP JOINT
Gluteus medius (GMed)
   Gluteus minimus
   Tensor fasciae latae

INTERNAL ROTATION OF THE HIP JOINT
Gluteus minimus (GMin)
   Gluteus medius
   Tensor fasciae latae

EXTERNAL ROTATION OF THE HIP JOINT
Piriformis
Obturator internus
Obturator externus
Gemellus superior
Gemellus inferior
Quadratus femoris

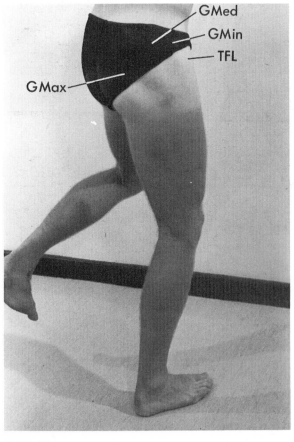

Fig. 4.1

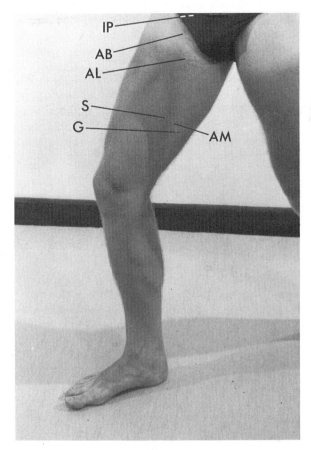

Fig. 4.2

GLUTEUS MAXIMUS

Attachments: Extends from the ilium behind the posterior gluteal line, the postero-lateral surface of the sacrum and coccyx and the sacrotuberous ligament, to the gluteal tuberosity of the femur (deep fibres) and the iliotibial band.
Nerve supply: Inferior gluteal L5, S1, 2.

Surface markings: It forms the major bulk of the buttock and can be palpated between its attachments.

Actions
Rostral end fixed: It extends and laterally rotates the hip joint. The lower fibres may assist in adduction and the upper fibres in abduction.
Caudal end fixed: It will assist in extending the trunk on the femur, as in shifting from four point to two point kneeling.

Maximum extensibility: The hip joint is flexed and internally rotated.

TESTING POSITION
Grades 5, 4 and 3 (Fig. 4.3): In prone lying, with the knee joint flexed to at least 90°, the hip is extended. Resistance is applied to the lower third of the thigh in the direction of flexion.
Grades 2, 1 and 0 (Fig. 4.4): In side lying, with the trunk stabilized in the mid position, the leg to be tested uppermost and supported with the thigh in a horizontal position and the knee joint in flexion, the hip is extended. The other leg is flexed at the hip and knee joints to assist in pelvic stabilization.

Comments: The position for Grades 5, 4 and 3 tests only the inner range.

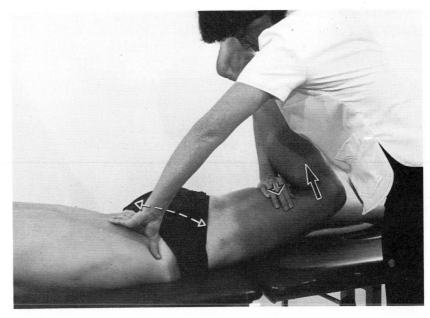

Fig. 4.3

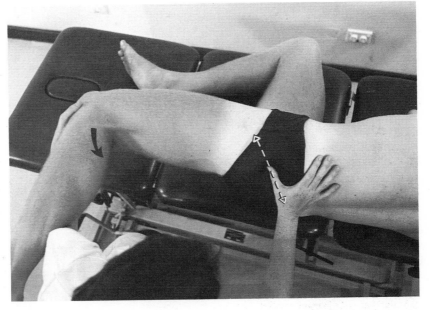

Fig. 4.4

Lower extremity

Muscles tested at the hip joint

PSOAS MAJOR

Attachments: Extends from the anterior
surfaces of the transverse processes, the
bodies and intervertebral discs of the
lumbar vertebrae to the lesser trochanter
of the femur.
Nerve supply: Lumbar plexus L1, 2, 3.

Surface markings: It can be palpated
immediately below the inguinal ligament
on the medial side of the sartorius.

ILIACUS

Attachments: Extends from the superior
two-thirds of the iliac fossa and the
internal surface of the iliac crest to the
lateral side of the psoas major tendon.
Nerve supply: Femoral L2, 3.

Surface markings: Too deep to be
palpated.

For psoas and iliacus (iliopsoas)

Actions
Rostral end fixed: These muscles flex the
hip joint by bringing the femur towards
the trunk.
Caudal end fixed: These muscles flex the
hip joint by bringing the trunk towards
the femur, as in performing sit-ups.

Maximum extensibility: In prone lying,
with the pelvis stabilized, the hip joint is
extended.

TESTING POSITION
Grades 5, 4 and 3 (Fig. 4.5): In supine
lying, with the knees flexed over the end
of the plinth and the opposite iliac crest
fixed, the femur is flexed towards the
trunk. Resistance is applied to the lower
end of the thigh in the direction of
extension.
Grades 2, 1 and 0 (Fig. 4.6): In side lying,
with the trunk stabilized in the mid
position, the leg to be tested uppermost
and supported with the thigh in a
horizontal position and the knee joint in
flexion, the hip is flexed. The other leg is
flexed at the hip and knee joints to assist
in pelvic stabilization.

Comments: It is important to ensure that
the hip remains in a neutral position
throughout the testing so that

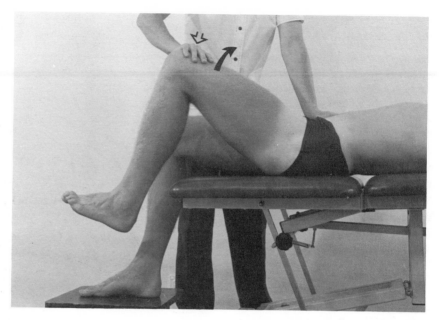

Fig. 4.5

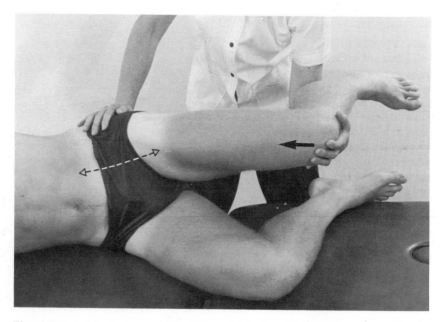

Fig. 4.6

substitution by sartorius is prevented.
Substitution by obliquus externus
abdominis will cause flattening of the
lumbar curve and may be mistaken for
iliopsoas activity. The position for
Grades 5, 4 and 3 only tests the middle
range. Inner range can be tested in the
sitting position.

TENSOR FASCIAE LATAE

Attachments: Extends from the anterior quarter of the iliac crest and the anterior superior iliac spine of the ilium to the iliotibial tract of the fasciae latae one-third the way down the tract.
Nerve supply: Superior gluteal L4, 5.

Surface markings: It can be palpated anterior to the greater trochanter.

Actions
Rostral end fixed: With the hip joint internally rotated, it flexes the hip. It assists also in abduction and internal rotation.
Caudal end fixed: It provides lateral stability to the knee joint, as in standing on one leg while kicking a ball.

Maximum extensibility: The hip is extended, laterally rotated and adducted. The knee is held in extension.

TESTING POSITION
Grades 5, 4 and 3 (Fig. 4.7): In side lying a quarter turn towards supine, the hip is flexed, abducted and internally rotated. Resistance is applied to the thigh in the direction of extension, adduction and external rotation.
Grades 2, 1 and 0 (Fig. 4.8): In side lying a quarter turn towards prone, the supported thigh is flexed and abducted.

Comments: The testing position chosen emphasises the tensor fascia latae as a hip joint flexor.

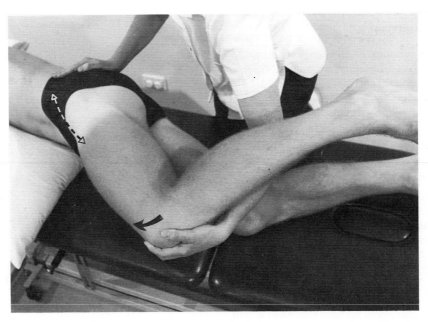

Fig. 4.7

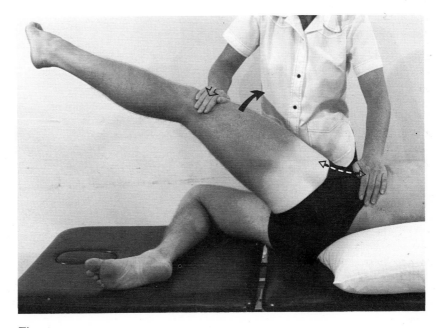

Fig. 4.8

Lower extremity

Muscles tested at the hip joint

SARTORIUS

Attachments: It extends from the anterior superior iliac spine to the medial surface of the proximal part of the tibial tuberosity.

Nerve supply: Femoral L2, 3.

Surface markings: It can be palpated just below the anterior superior iliac spine.

Actions

Rostral end fixed: It flexes, externally rotates and adducts the hip joint. It assists also in flexion of the knee and may internally rotate the tibia on the femur.

Caudal end fixed: It can assist in flexing the hip joint by bringing the trunk towards the femur, as in performing sit-ups. It also provides stability to the medial side of the knee joint when weight bearing on one leg, as in kicking a ball.

Maximum extensibility: The hip joint is extended, internally rotated and abducted with the knee joint extended.

TESTING POSITION

Grades 5, 4 and 3 (Fig. 4.9): In supine lying, with the knee joint supported in flexion and the pelvis stabilized, the hip is flexed, externally rotated and adducted. Resistance is applied to the medial surface of the knee in the direction of extension, internal rotation and abduction.

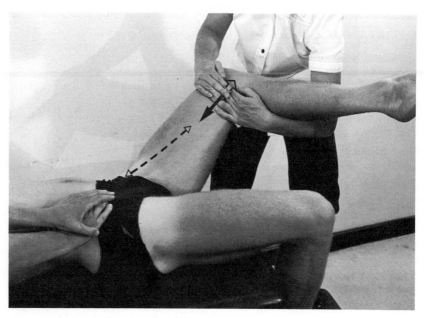

Fig. 4.9

Grades 2, 1 and 0 : A gravity eliminated position is not practical.

Comments: The hip joint must be maintained in external rotation to isolate the hip flexion action of sartorius.

ADDUCTOR MAGNUS

Attachments: Extends from the ischial tuberosity and ischiopubic ramus to the linea aspera and adductor tubercle of the femur.
Nerve supply: Obturator, L2, 3, 4 and Sciatic L4, 5.

Surface markings: It can be palpated on the medial side of the thigh close to its rostral attachment.

Actions
Rostral end fixed: It adducts the hip. The anterior fibres may assist in flexion and the posterior fibres in extension of the hip.
Caudal end fixed: It assists in stabilizing the pelvis and femur during the stance phase of gait.

ADDUCTOR LONGUS

Attachments: Extends from the anterior surface of the body of the pubis to the linear aspera of the femur.
Nerve supply: Obturator L2, 3, 4.

Surface markings: It can be palpated on the medial side of the thigh at the level of the inguinal ligament lateral to the tendon of adductor longus.

Actions
Rostral end fixed: It adducts the hip. It assists also in hip flexion.
Caudal end fixed: It assists in stabilizing the pelvis and femur during the stance phase of gait.

ADDUCTOR BREVIS

Attachments: Extends from the lateral surface of the inferior ramus of the pelvis to the medial surface of the upper third of the linea aspera.
Nerve supply: Obturator L2, 3, 4.

Surface markings: Too deep to palpate.

Actions
Rostral end fixed: It adducts the hip. It assists also in hip flexion.
Caudal end fixed: It assists in stabilizing the pelvis and femur during the stance phase of gait.

GRACILIS

Attachments: Extends from the ramus of the pubis near the symphysis to the medial surface of the shaft of the tibia just below the medial tuberosity.
Nerve supply: Obturator L2, 3.

Surface markings: Its tendon can be palpated on the medial surface of the thigh, at the level of the inguinal ligament, medial to that of adductor longus.

Actions
Rostral end fixed: It adducts the hip and also flexes and medially rotates it.
Caudal end fixed: It provides medial stability to the knee joint in weight bearing on one leg, as in kicking a ball.

PECTINEUS

Attachments: Extends from the pectineal line of the pubis to the femur along a line joining the lesser trochanter to the linea aspera.
Nerve supply: Femoral L2, 3 and Accessory Obturator L3.

Surface markings: It forms the floor of the femoral triangle and can be palpated on the medial surface of the thigh near its rostral attachment.

Actions
Rostral end fixed: It adducts the hip and assists also in hip flexion.
Caudal end fixed: It assists in stabilizing the pelvis and femur during the stance phase of gait.

(contd)

Lower extremity

Muscles tested at the hip joint

For all the hip adductors

Maximum extensibility: With one leg fixed in full abduction, the opposite hip is abducted and externally rotated.

TESTING POSITION
Grades 5, 4 and 3 (Fig. 4.10): In side lying, with the leg not being tested supported in abduction and neutral rotation, the other leg is adducted. There must be no flexion, extension or rotation of the hip and no movement of the pelvis. Resistance is applied to the medial side of the knee joint in the direction of abduction.
Grades 2, 1 and 0 (Fig. 4.11): In supine lying, with the knee of the leg not being tested flexed over the side of the supporting surface to stabilize the pelvis, the supported leg is adducted.

Comments: During the Grades 5, 4 and 3 muscle test, it is essential that the angle of abduction of the uppermost leg is maintained throughout the test.

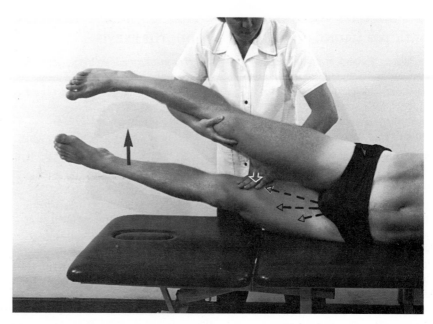

Fig. 4.10

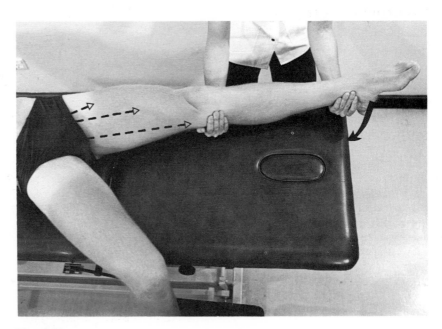

Fig. 4.11

GLUTEUS MEDIUS

Attachments: Extends from the lateral surface of the ilium between the iliac crest and the posterior gluteal line to the lateral surface of the greater trochanter of the femur.

Nerve supply: Superior gluteal L5, S1.

Surface markings: It can be palpated just above the greater trochanter.

Actions

Rostral end fixed: It abducts the hip. The anterior fibres may assist in flexion and internal rotation and the posterior fibres in extension and lateral rotation of the hip.

Caudal end fixed: It holds the pelvis horizontal during single leg support.

Maximum extensibility: The hip joint is adducted, extended and externally rotated for the anterior fibres and adducted, flexed and internally rotated for the posterior fibres.

TESTING POSITION

Grades 5, 4 and 3 (Fig. 4.12): In side lying, with the leg being tested uppermost, and the other leg flexed at both the hip and knee joints to stabilize the pelvis, the uppermost leg is abducted and slightly extended. Resistance is applied to the lateral surface of the knee joint in the direction of adduction and slight flexion.

Grades 2, 1 and 0 (Fig. 4.13): In supine lying, with the knee of the leg not being tested flexed over the side of the supporting surface to stabilize the pelvis, the supported leg is abducted.

Comments: Weakness of gluteus medius causes the Trendelenburg sign seen in single leg stance. Figure 4.48, page **114** demonstrates this effect.

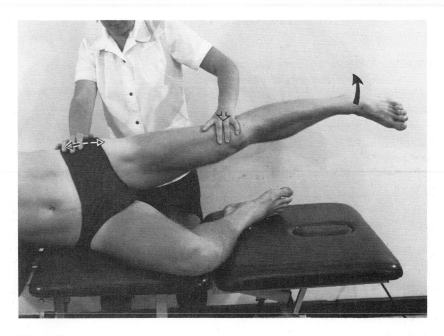

Fig. 4.12

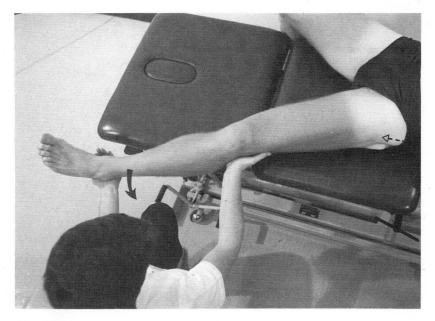

Fig. 4.13

Lower extremity

Muscles tested at the hip joint

GLUTEUS MINIMUS

Attachments: It extends from the lateral surface of the ilium between the middle and inferior gluteal lines, to the greater trochanter of the femur.
Nerve Supply: Superior gluteal L5, S1.

Surface markings: It is too deep to palpate.

Actions
Rostral end fixed: It internally rotates the hip and also assists in abduction.
Caudal end fixed: It assists gluteus medius to hold the pelvis in the horizontal position during single leg support.

Maximum extensibility: The hip joint is externally rotated and adducted.

TESTING POSITION
Grades 5, 4 and 3 (Fig. 4.14): In sitting, with the knees flexed over the edge of the supporting surface and the thigh in a neutral position between abduction and adduction, the hip is internally rotated (the heel is moved away from the other leg). Resistance is applied to the lateral surface of the ankle in the direction of external rotation. The thigh must be stabilized during the test.
Grades 2, 1 and 0 (Fig. 4.15): In supine lying, with the hip and knee flexed to 90°, the lower leg supported, the hip is internally rotated. The thigh is stabilized in a vertical position.

Comments: In the testing position for Grades 5, 4 and 3, the outer range of the movement is not against gravity. It is not practical to test this range against gravity. In the testing position for Grades 2, 1 and 0, the thigh must remain vertical, otherwise abduction of the hip may be substituted for the rotation.

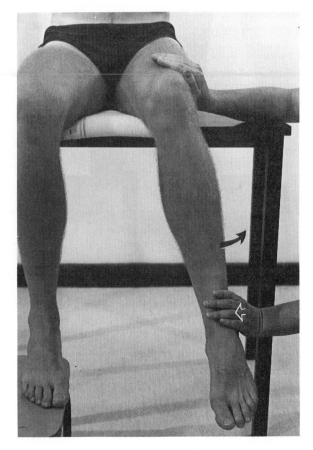

Fig. 4.14

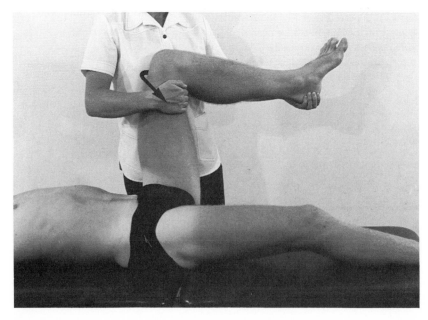

Fig. 4.15

PIRIFORMIS

Attachments: Extends from the anterior surface of the sacrum, lateral to the first four foramina and the margin of the greater sciatic notch, and passes through the greater sciatic notch to the superior surface of the greater trochanter of the femur.
Nerve supply: L5, S1, 2.

OBTURATOR INTERNUS

Attachments: Extends from the internal surface of the obturator membrane and the margin of the obturator foramen, and passes through the lesser sciatic notch to the medial surface of the greater trochanter of the femur.
Nerve supply: Sacral plexus L5, S1.

OBTURATOR EXTERNUS

Attachments: Extends from the obturator membrane and the rami of the pubis and ischium to the trochanteric fossa of the femur.
Nerve supply: Obturator L3, 4.

GEMELLUS SUPERIOR AND INFERIOR

Attachments: These muscles extend from the ischial spine and the margin of the lesser sciatic notch to the medial surface of the greater trochanter of the femur via the obturator tendon.
Nerve supply: Sacral plexus L5, S1.

QUADRATUS FEMORIS

Attachments: It extends from the lateral border of the proximal part of the ischial tuberosity to the intertrochanteric ridge of the femur.
Nerve supply: Sacral plexus L5, S1.

(contd)

Lower extremity

Muscles tested at the hip joint

For all the external rotators

Surface markings: These muscles are too deep to be palpated directly, but some identification may be possible when the gluteus maximus is fully relaxed.

Actions

Rostral end fixed: These muscles externally rotate the hip. When the hip is in flexion, they assist in abduction except for obturator externus, which assists in adduction. Piriformis also assists in extension of the hip joint.

Caudal end fixed: These muscles assist in stabilizing the pelvis during single leg support.

Maximum extensibility: With the hip and knee joints flexed to 90° the hip is internally rotated.

TESTING POSITION

Grades 5, 4 and 3 (Fig. 4.16): In sitting, with the knees flexed over the edge of the supporting surface and the thigh in a neutral position between abduction and adduction, the hip is externally rotated (the heel is moved towards the other leg). Resistance is applied to the medial surface of the ankle in the direction of internal rotation. The thigh must be fixed during the test.

Grades 2, 1 and 0 (Fig. 4.17): In supine lying, with the hip and knee flexed to 90° and the lower leg supported, the hip is internally rotated. The thigh is stabilized in a vertical position.

Comments: In the testing position for Grades 5, 4 and 3, the outer range of the movement is not against gravity. It is not practical to test this range against gravity. In the testing position for Grades 2, 1 and 0, the thigh must remain vertical, otherwise abduction of the hip may be substituted for the rotation.

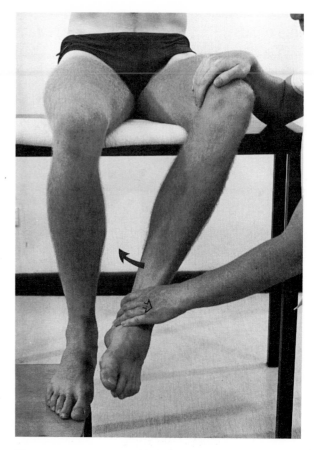

Fig. 4.16

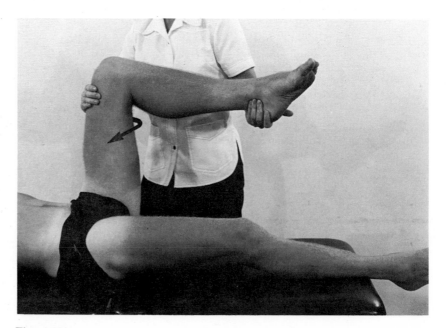

Fig. 4.17

Muscles tested at the knee joint (Figs 4.18 & 4.19)

FLEXION OF THE KNEE
   Biceps femoris (BF)   ⎫ Hamstring
   Semitendinosus (ST)   ⎬ group.
   Semimembranosus (SM) ⎭

EXTENSION OF THE KNEE
  Quadriceps femoris:
    Rectus femoris (RF)
    Vastus medialis (VM)
    Vastus lateralis
    Vastus intermedius
  Popliteus

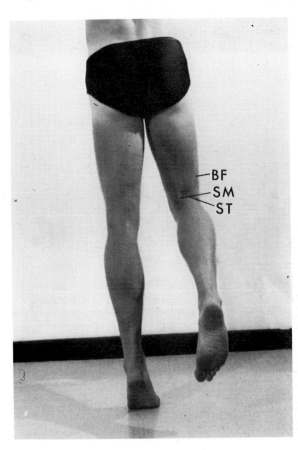

Fig. 4.18

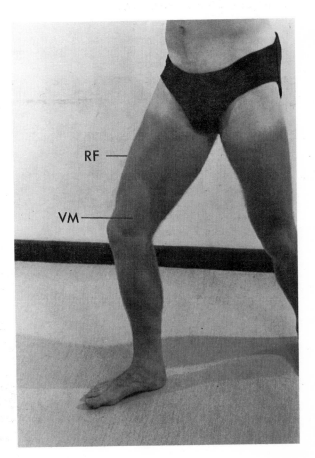

Fig. 4.19

Muscles tested at the knee joint

BICEPS FEMORIS

Attachments: The long head originates from the ischial tuberosity, and the short head from the lower half of the lateral lip of the linea aspera. Both heads insert into the head of the fibula.

Nerve supply: Sciatic (tibial branch) L5, S1, 2 to the long head and the Sciatic (peroneal branch) L5, S1, 2 to the short head.

Surface markings: The tendon can be palpated on the lateral side of the knee just above the head of the fibula.

Actions

Rostral end fixed: It flexes and laterally rotates the knee joint, such that the tibia moves towards the thigh.

Caudal end fixed: The long head assists in extension of the hip and the lateral head in internal rotation of the hip joint, during activities such as walking and running.

SEMITENDINOSUS

Attachments: It extends from the ischial tuberosity to the medial surface of the shaft of the tibia just below the tibial condyle.

Nerve supply: Sciatic (tibial branch) L5, S1, 2.

Surface markings: The tendon can be palpated on the medial side of the knee where it is lateral to that of semimembranosus.

Actions

Rostral end fixed: It flexes and medially rotates the knee joint, such that the tibia moves towards the thigh.

Caudal end fixed: It assists with extension and internal rotation of the hip joint, during activities such as walking and running.

SEMIMEMBRANOSUS

Attachments: Extends from the ischial tuberosity to the medial condyle of the tibia.

Nerve supply: Sciatic (tibial branch) L5, S1, 2.

Surface markings: The tendon can be palpated on the medial side of the knee, in front of that of semitendinosus.

Actions

Rostral end fixed: It flexes the knee joint and internally rotates the tibia.

Caudal end fixed: It assists with extension and medial rotation of the hip joint.

For all the knee flexors

Maximum extensibility: With the hip joint flexed to approximately 90°, the knee is extended and the hip joint laterally rotated (semimembranosus and semitendinosus), or medially rotated (biceps femoris).

TESTING POSITION

Grades 5, 4 and 3 (Fig. 4.20): In prone lying, with the hip joint in neutral rotation and adduction, the knee joint is flexed. Resistance is applied to the posterior surface of the tibia in the direction of extension. The pelvis is stabilized over the posterior part of the iliac crest.

Grades 2, 1 and 0 (Fig. 4.21): In side lying, with the trunk stabilized in the mid position, the leg to be tested uppermost and supported with the thigh in a horizontal position, the knee joint is flexed. The other leg is flexed at the hip and knee joints to assist in pelvic stabilization.

Comments: Although the test for this muscle group requires that the hip be in neutral rotation, it is advisable to check the contribution of the medial and lateral components to the movement. This is done by placing the hip in external rotation and then eliciting knee flexion (to make the biceps femoris most active) or placing the hip in internal rotation and eliciting knee flexion (to make the semimembranosus and semitendinosus active).

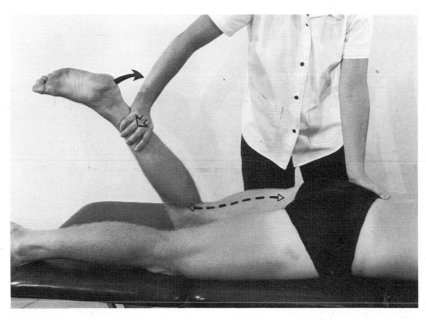

Fig. 4.20

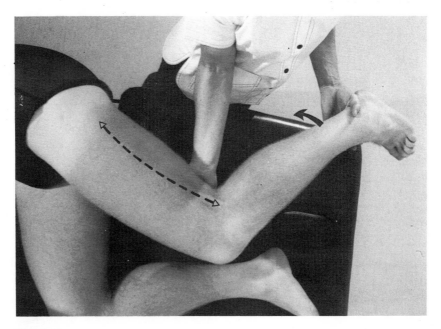

Fig. 4.21

Lower extremity

Muscles tested at the knee joint

QUADRICEPS FEMORIS

Attachments: The rectus femoris originates from the anterior inferior spine of the ilium and the upper margin of the acetabulum. The vastus medialis originates from the medial line of the linea aspera. The vastus intermedius originates from the upper three-quarters of the shaft of the femur. The vastus lateralis originates from the lateral line of the linea aspera. All four heads insert into the upper surface of the patella and then via the patella ligament into the tibial tuberosity.
Nerve supply: Femoral L2, 3, 4.

Surface markings: All components except the vastus intermedius are superficial and can be readily palpated on the anterior aspect of the thigh.

Actions
Rostral end fixed: It extends the knee joint. The rectus femoris also flexes the hip joint.
Caudal end fixed: It controls flexion of the knee when weight bearing, as in squatting or descending stairs.

Maximum extensibility: The knee joint is flexed. However, to examine rectus femoris, the knee joint must be flexed with the hip joint in extension.

TESTING POSITION
Grades 5, 4 and 3 (Fig. 4.22): In sitting with the lumbar lordosis maintained and the knees flexed over the edge of the supporting surface, the knee is extended. Resistance is applied to the anterior surface of the lower end of the leg in the direction of flexion. The thigh is stabilized.
Grades 2, 1 and 0 (Fig. 4.23): In side lying with the trunk stabilized in the mid position, the leg to be tested uppermost, supported with the thigh in a horizontal position and the knee joint in flexion, the knee joint is extended. The other leg is flexed at the hip and knee joints to assist in pelvic stabilization.

Comments: The principal role of the muscle is control of the knee in weight bearing. Figure 4.49, page **115** shows the way in which this functional component of the muscle should be evaluated.

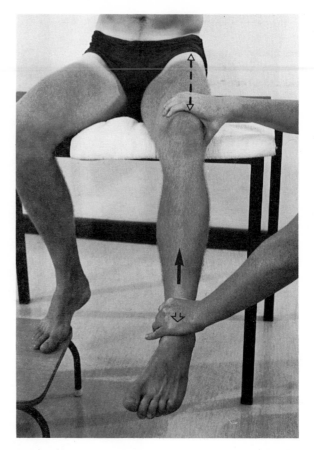

Fig. 4.22

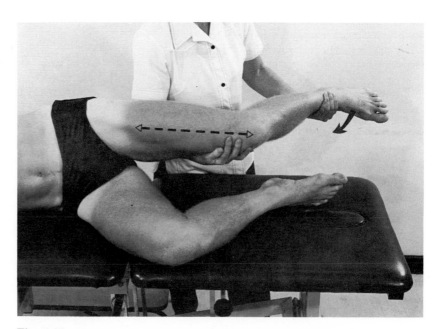

Fig. 4.23

POPLITEUS

Attachments: Extends from the lateral condyle of the femur to the posterior surface of the tibia just above the oblique line.
Nerve supply: Tibial L4, 5, S1.

Surface markings: The tendon can be palpated postero-laterally in the popliteal fossa.

Actions
Rostral end fixed: It internally rotates the tibia on the femur.
Caudal end fixed: It externally rotates the femur on the tibia and flexes the knee, as in squatting.

Maximum extensibility: With the tibia externally rotated on the femur, the knee is extended.

TESTING POSITION
Grades 5, 4 and 3: As for quadriceps femoris (above).
Grades 2, 1 and 0 (Not illustrated): In sitting, with the tibia laterally rotated over the edge of the supporting surface, the tibia is medially rotated.

Comments: This muscle is usually tested with the quadriceps muscle group.

Lower extremity

Muscles tested at the ankle joint (Figs 4.24 & 4.25)

PLANTARFLEXORS OF THE
ANKLE
Soleus (S)
Gastrocnemius (G)
Plantaris
Tibialis posterior

DORSIFLEXORS OF THE ANKLE
Tibialis anterior (TA)
Peroneus tertius
Extensor digitorum longus
Extensor hallucis longus

INVERTORS OF THE ANKLE
Tibialis posterior (TP)
Tibialis anterior

EVERTORS OF THE ANKLE
Peroneus longus (PL)
Peroneus brevis (PB)
Peroneus tertius

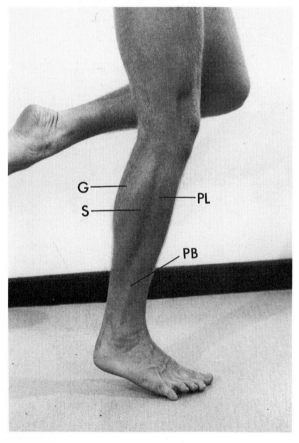

Fig. 4.24

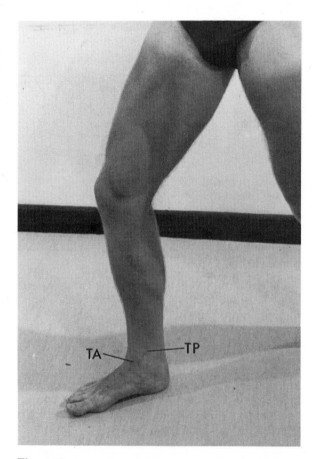

Fig. 4.25

SOLEUS

Attachments: Extends from the posterior surface of the head and shaft of the fibula, and from the middle one-third of the medial border of the tibia plus the tendinous arch between the tibia and fibula. It inserts with the gastrocnemius into the calcaneum via the tendocalcaneum.

Nerve supply: Tibial S1, 2.

Surface markings: It can be palpated on the posterior surface of the calf on either side of the tendocalcaneum.

Actions

Rostral end fixed: It plantarflexes the ankle joint.

Caudal end fixed: It raises the heel from the supporting surface so that weight is taken through the ball of the foot, as in standing on tip-toe, or preparing to jump.

Maximum extensibility: With the knee flexed the ankle is dorsiflexed.

TESTING POSITION

Grades 5, 4 and 3 (Fig. 4.26): In prone lying, with the knee flexed to 90°, the ankle is plantarflexed. Resistance is applied to the ball of the foot in the direction of dorsiflexion. The tibia is stabilized near the ankle.

Grades 2, 1 and 0 (Not illustrated): In side lying, with the leg to be tested uppermost and the knee joint flexed to 90°, the ankle is plantarflexed.

Comments: This muscle usually acts against gravity and the full weight of the body. The functional test is described with gastrocnemius in Figure 4.50, on page **115**.

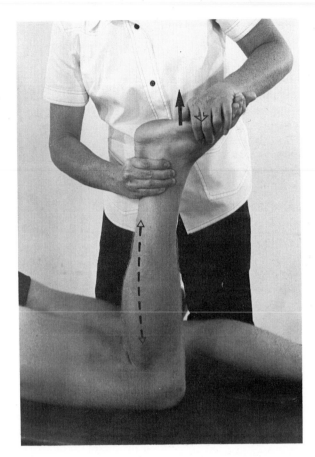

Fig. 4.26

Muscles tested at the ankle joint

GASTROCNEMIUS

Attachments: The lateral and medial heads originate from the lateral and medial condyles of the femur and insert, with the soleus, into the posterior surface of the calcaneum via the tendocalcaneum.
Nerve supply: Tibial S1, 2.

Surface markings: It can be palpated on either side of the posterior surface of the calf, just below the knee joint.

PLANTARIS

Attachments: Extends from the lateral supracondylar ridge of the femur to the medial margin of the tendocalcaneum and plantar fascia of the foot.
Nerve supply: Tibial S1, 2.

Surface markings: Too deep to be palpated.

For gastrocnemius and plantaris

Actions
Rostral end fixed: These muscles plantarflex the ankle and also assist in flexion of the knee joint.
Caudal end fixed: They raise the heel from the supporting surface so that weight is taken on the ball of the foot, as when standing on tip-toe, or preparing to jump.

Maximum extensibility: With the knee extended, the foot is dorsiflexed.

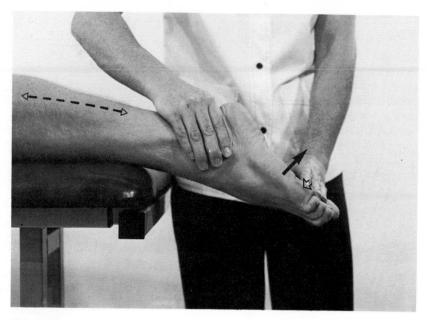

Fig. 4.27

TESTING POSITION
Grades 5, 4 and 3 (Fig. 4.27): In prone lying with the knee extended, the ankle is plantarflexed. Resistance is applied to the ball of the foot in the direction of dorsiflexion.
Grades 2, 1 and 0 (Not illustrated): In side lying, with the leg to be tested uppermost and the knee joint extended, the ankle is plantarflexed.

Comments: These muscles usually act against gravity and the full weight of the body. The functional test is shown in Figure 4.50, page **115**.

TIBIALIS ANTERIOR

Attachments: Extends from the upper two-thirds of the antero-lateral surface of the tibia and interosseous membrane to the base of the first metatarsal and medial cuneiform bones.

Nerve supply: Deep peroneal L4, 5.

Surface markings: The tendon can be palpated on the anterior surface of the ankle close to its caudal attachment.

Actions

Rostral end fixed: It dorsiflexes the ankle and inverts the forefoot.

Caudal end fixed: It brings the tibia forward over the foot, as occurs in the mid-stance phase of walking or running.

Maximum extensibility: The ankle joint is plantarflexed and the forefoot everted.

TESTING POSITION

Grades 5, 4 and 3 (Fig. 4.28): In sitting, with the knee flexed over the edge of the supporting surface, the ankle joint is dorsiflexed and the forefoot inverted. Resistance is applied to the medial and anterior surface of the foot in the direction of plantarflexion and eversion.

Grades 2, 1 and 0 (Fig. 4.29): In supine or half lying, with the hip joint internally rotated and the knee joint slightly flexed, the ankle joint is dorsiflexed and the forefoot inverted.

Comments: When tibialis anterior is weak, the dorsiflexion action of extensor hallucis, extensor digitorum longus and peroneus tertius may produce the movement.

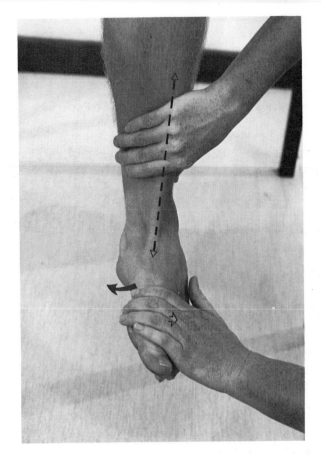

Fig. 4.28

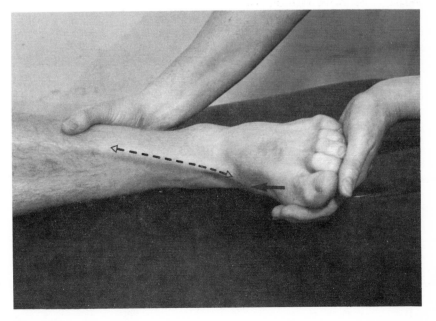

Fig. 4.29

Lower extremity

Muscles tested at the ankle joint

PERONEUS TERTIUS

Attachments: Extends from the lower third of the anterior surface of the fibula and interosseus membrane to the medial part of the dorsal surface of the base of the fifth metatarsal bone.
Nerve supply: Deep peroneal L5, S1.

Surface markings: The tendon can be palpated on the lateral surface of the foot as it inserts into the fifth metatarsal. It is difficult to distinguish from the tendon of peroneus brevis.

Actions
Rostral end fixed: It dorsiflexes the foot, acting as part of extensor digitorum longus. It may assist in eversion also.
Caudal end fixed: It stabilizes the ankle joint during single leg support, for example in hurdling or executing the hop, step and jump.

Maximum extensibility: With the ankle joint plantarflexed and the forefoot inverted, the lateral four toes are flexed.

TESTING POSITION
Grades 5, 4 and 3 (Not illustrated): In sitting, with the knee flexed over the edge of the supporting surface and the forefoot everted, the ankle joint is dorsiflexed. Resistance is applied to the lateral and anterior surface of the foot in the direction of plantarflexion.
Grades 2, 1 and 0 (Not illustrated): In side lying, with the leg to be tested uppermost, the knee slightly flexed and the forefoot everted, the ankle joint is dorsiflexed.

Comments: The dorsiflexion action of extensor digitorum longus is likely to be apparent in the test since peroneus tertius is intimately related to this muscle.

TIBIALIS POSTERIOR

Attachments: Extends from the posterior surface of the shafts of the tibia and fibula to the tuberosity of the navicular and by fibrous tissue to the three cuneiform and cuboid bones and to the sustentaculum tali and base of the second, third, and fourth metatarsal bones.

Nerve supply: Tibial L4, 5.

Surface markings: The tendon can be palpated as it passes immediately behind the medial malleolus.

Actions

Rostral end fixed: It plantarflexes and inverts the foot.

Caudal end fixed: It stabilizes the ankle in single leg support, for example in performing the hop, step and jump.

Maximum extensibility: The ankle joint is dorsiflexed and the forefoot everted.

TESTING POSITION

Grades 5, 4 and 3 (Fig. 4.30): In sitting, with the knee flexed over the supporting surface or in prone lying with the knee flexed to 90°, the ankle is plantarflexed and the forefoot inverted. Resistance is applied to the medial surface and ball of the foot in the direction of dorsiflexion and eversion.

Grades 2, 1 and 0 (Fig. 4.31): In supine or half lying, with the knee slightly flexed, the ankle is plantarflexed and the forefoot inverted.

Comments: Grades 5, 4 and 3 are not against gravity.

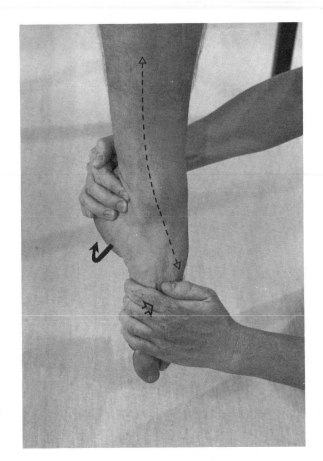

Fig. 4.30

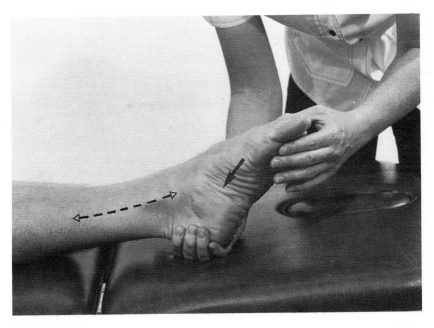

Fig. 4.31

Muscles tested at the ankle joint

PERONEUS LONGUS

Attachments: Extends from the upper two-thirds of the lateral surface of the fibula, passes behind the lateral malleolus and crosses the sole of the foot to the base of the first metatarsal and medial cuneiform bone.

Nerve supply: Superficial peroneal L5, S1, 2.

Surface markings: The tendon can be palpated as it passes behind the lateral malleolus.

Actions

Rostral end fixed: It everts the forefoot and depresses the first metatarsal. It assists in dorsiflexion of the ankle joint.

Caudal end fixed: It stabilizes the ankle joint during single leg support, for example in hurdling.

Fig. 4.32

PERONEUS BREVIS

Attachments: Extends from the lower two-thirds of the lateral surface of the fibula, passes behind the lateral malleolus anterior to peroneus longus, and attaches to the lateral side of the tubercle on the base of the fifth metatarsal bone.

Nerve supply: Superficial peroneal L5, S1, 2.

Surface markings: The tendon can be palpated on the lateral surface of the foot as it inserts into the fifth metatarsal.

Actions

Rostral end fixed: It everts the forefoot.

Caudal end fixed: It stabilizes the ankle joint during single leg support.

For peroneus longus and brevis

Maximum extensibility: The forefoot is inverted and the ankle joint is plantarflexed.

TESTING POSITION

Grades 5, 4 and 3 (Fig. 4.32): In sitting, with the knee joint flexed over the edge of the supporting surface, the forefoot is everted. Resistance is applied to the lateral and anterior surface of the foot in the direction of inversion.

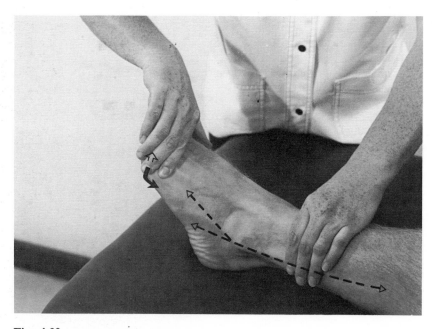

Fig. 4.33

Grades 2, 1 and 0 (Fig. 4.33): In supine or half lying, with the knee slightly flexed, the forefoot is everted.

Comments: The tendons of each of the muscles should be palpated to detect their contribution to the test movement.

Muscles tested at the joints of the toes (Fig. 4.34)

EXTENSION OF THE GREAT TOE
Extensor hallucis longus (EHL)
Extensor hallucis brevis

EXTENSION OF THE LATERAL FOUR TOES
Extensor digitorum longus (EDL)
Extensor digitorum brevis (EDB)

FLEXION OF THE GREAT TOE
Flexor hallucis longus
Flexor hallucis brevis

FLEXION OF THE LATERAL FOUR TOES
Flexor digitorum longus
Flexor digitorum brevis
Flexor digitorum accessorius

ABDUCTION OF THE TOES
Abductor hallucis
Abductor digiti minimi
Dorsal interossei

ADDUCTION OF THE TOES
Adductor hallucis
Plantar interossei

FLEXION OF THE METATARSOPHALANGEAL JOINTS
Lumbricals

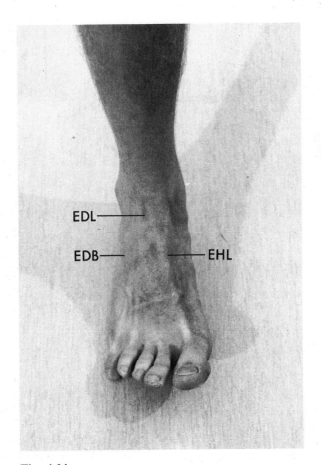

Fig. 4.34

Muscles tested at the joints of the toes

EXTENSOR HALLUCIS LONGUS

Attachments: Extends from the anterior surface of the middle half of the fibula and interosseous membrane to the base of the distal phalanx of the great toe.
Nerve supply: Deep peroneal L5, S1.

Surface markings: The tendon can be palpated between the medial and lateral malleoli just lateral to the tendon of tibialis anterior or over the dorsal surface of the proximal phalanx of the great toe.

Actions
Rostral end fixed: Extends all the joints of the great toe. It assists also in dorsiflexion of the ankle joint and inversion of the forefoot.
Caudal end fixed: It assists in stabilizing the great toe and ankle joint during weight bearing in the 'en pointe' position.

Maximum extensibility: With the ankle joint plantarflexed, the forefoot everted, all the joints of the great toe are flexed.

TESTING POSITION
Grades 5, 4 and 3 (Fig. 4.35): In supine lying or long sitting, with the supported foot in a plantargrade position, the great toe is extended. Resistance is applied to the distal phalanx of the great toe in the direction of flexion.
Grades 2, 1 and 0: In side lying, with the leg to be tested uppermost and supported, the distal phalanx of the great toe is extended. When weakness is present, proximal fixation will be required.

Comments: Figure 4.36 demonstrates the Grades 2, 1 and 0 position of testing for all flexor and extensor muscles of the toes.

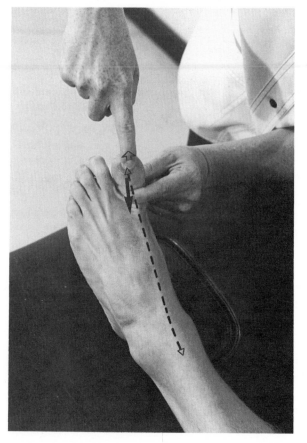

Fig. 4.35

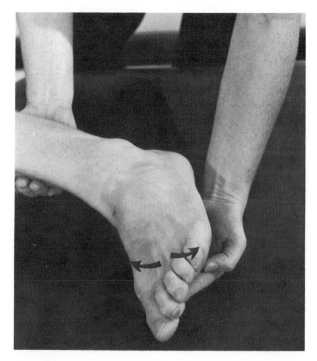

Fig. 4.36

EXTENSOR HALLUCIS BREVIS

Attachments: Extends from the lateral and upper surfaces of the distal part of the calcaneus, the lateral talocalcaneal ligament and the top of the inferior extensor retinaculum, to the dorsal surface of the base of the proximal phalanx of the great toe.
Nerve supply: Deep peroneal S1, 2.

Surface markings: The tendon, which lies lateral to that of extensor hallucis longus, can be palpated as it approaches the base of the proximal phalanx of the great toe.

Actions
Rostral end fixed: It extends the metatarsophalangeal joint of the great toe.
Caudal end fixed: It assists in stabilizing the great toe during weight bearing in the 'en pointe' position.

Maximum extensibility: With the foot in the plantargrade position the great toe is flexed.

TESTING POSITION
Grades 5, 4 and 3 (Not illustrated): In supine lying or long sitting, with the foot supported in a plantargrade position the great toe is extended. Resistance is applied to the proximal phalanx of the great toe in the direction of flexion.
Grades 2, 1 and 0 (Same position as Fig. 4.36): In side lying, with the leg to be tested uppermost and supported, the metatarsophalangeal joint of the great toe is extended.

Comments: The testing position for Grades 5, 4 and 3 is the same as that shown in Figure 4.35. However, the resistance is applied over the proximal phalanx.

Lower extremity

Muscles tested at the joints of the toes

EXTENSOR DIGITORUM LONGUS

Attachments: Extends from the upper two-thirds of the anterior surface of the fibula and inserts by four tendons into the dorsal surface of the four lateral toes via an extensor expansion. The intermediate slip inserts into the base of the middle phalanx, and the two lateral slips into the base of the distal phalanx.
Nerve supply: Deep peroneal L5, S1.

Surface markings: The tendons can be palpated on the dorsal surface of the metatarsophalangeal joints of the lateral four toes.

Actions
Rostral end fixed: It extends the four lateral toes. It also assists in dorsiflexion of the ankle joint and eversion of the forefoot.
Caudal end fixed: It stabilizes the ankle joint during weight bearing in the 'en pointe' position.

EXTENSOR DIGITORUM BREVIS

Attachments: Extends from the lateral and upper surfaces of the distal part of the calcaneus, the lateral talocalcaneal ligament and the top of the inferior extensor retinaculum, via four tendons, to the lateral side of the tendons of extensor digitorum longus of the first, second, third and fourth digits. (The most medial slip forms the extensor hallucis brevis.)
Nerve supply: Deep peroneal S1, 2.

Surface markings: It can be palpated on the dorsum of the foot just distal to the lateral malleolus and lateral to the extensor digitorum tendon of the little toe.

Actions
Rostral end fixed: It extends the metatarsophalangeal joints of the four medial toes and assists in extending the interphalangeal joints of the same four toes.
Caudal end fixed: It assists in stabilizing the toes when weight is taken in the 'en pointe' position of the foot.

For extensor digitorum longus and brevis

Maximum extensibility: With the ankle joint plantarflexed and the forefoot inverted, the lateral four toes are flexed.

TESTING POSITION
Grades 5, 4 and 3 (Fig. 4.37): In supine lying or long sitting, with the foot supported in the plantargrade position, the lateral four toes are extended. Resistance is applied to the dorsal surface of the distal ends of the lateral four toes in the direction of flexion.
Grades 2, 1 and 0 (Same position as for Fig. 4.36, p. **104**): In side lying, with the leg to be tested uppermost and supported, the lateral four toes are extended.

Comments: In the presence of weakness, stabilization of the forefoot is essential to isolate the movement. This stabilization is not shown in Figure 4.36.

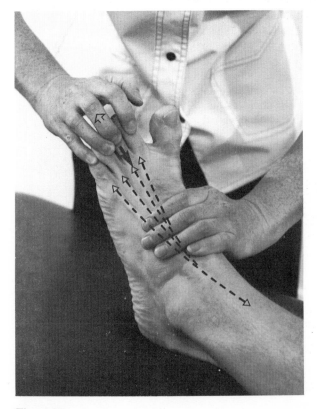

Fig. 4.37

FLEXOR HALLUCIS LONGUS

Attachments: Extends from the posterior surface of the lower two-thirds of the fibula and adjacent interosseous membrane to the base of the distal phalanx of the great toe.
Nerve supply: Tibial, S2, 3.

Surface markings: The tendon can be palpated on the plantar surface of the proximal phalanx of the great toe.

Actions
Rostral end fixed: It flexes all joints of the great toe. It also assists in plantarflexion of the ankle joint and inversion of the forefoot.
Caudal end fixed: It assists in stabilizing the great toe during weight bearing in the 'en pointe' position.

Maximum extensibility: With the ankle joint dorsiflexed and the forefoot everted, the great toe is extended.

TESTING POSITION
Grades 5, 4 and 3 (Fig. 4.38): In supine lying or long sitting, with the supported foot in a plantargrade position, the distal interphalangeal joint of the great toe is flexed. Resistance is applied to the plantar surface of the distal phalanx in the direction of extension.

Grades 2, 1 and 0 (Same position as Fig. 4.36, p. **104**): In side lying, with the leg to be tested uppermost and supported, the distal phalanx of the great toe is flexed.

Comments: The testing position for Grades 5, 4, and 3 has gravity assisting the movement, but because the limb segment being moved is so small the effect of gravity on the movement is considered minimal. It is more important that the subject is able to see the great toe so that vision can assist the patient to complete the movement. When weakness is present, proximal fixation will be required.

FLEXOR HALLUCIS BREVIS

Attachments: Extends from the medial surface of the cuboid and the middle and lateral cuneiform bones to either side of the base of the proximal phalanx of the great toe.
Nerve supply: Medial plantar S2, 3.

Surface markings: The tendon can be palpated just proximal to the metatarsophalangeal joint of the great toe.

Actions
Rostral end fixed: It flexes the metatarsophalangeal joint of the great toe.
Caudal end fixed: It assists with the stability of the great toe during weight bearing in the 'en pointe' position.

Maximum extensibility: With the foot in a plantargrade position, the metatarsophalangeal joint of the great toe is extended.

TESTING POSITION
Grades 5, 4 and 3 (Fig. 4.39): In supine lying or long sitting, with the supported foot in a plantargrade position, the metatarsophalangeal joint of the great toe is flexed. Resistance is applied to the plantar surface of the proximal phalanx of the great toe in the direction of extension.
Grades 2, 1 and 0 (Same position as Fig. 4.36, p. **104**): In side lying, with the leg to be tested uppermost and supported, the metatarsophalangeal joint of the great toe is flexed.

Comments: When weakness is present, proximal fixation will be required.

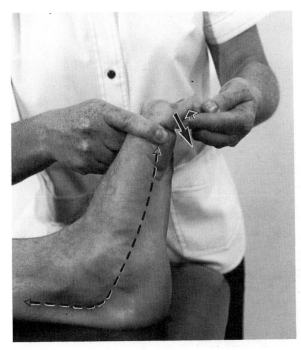

Fig. 4.38

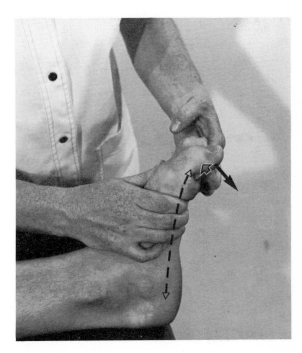

Fig. 4.39

Lower extremity

Muscles tested at the joints of the toes

FLEXOR DIGITORUM LONGUS

Attachments: Attaches to the middle third of the posterior surface of the tibia and extends, by four tendons, to the plantar surface of the distal phalanges of the lateral four toes.
Nerve supply: Tibial, S2, 3.

Surface markings: The tendons can be palpated on the plantar surface of the proximal interphalangeal joints of the lateral four toes. It is difficult to differentiate between the tendons of this muscle and those of the flexor digitorum brevis.

Actions
Rostral end fixed: It flexes the metatarsophalangeal joints and the interphalangeal joints of the lateral four toes. It also assists in plantarflexion of the ankle joint and inversion of the forefoot.
Caudal end fixed: It assists in stabilizing the ankle joint and forefoot during weight bearing in the 'en pointe' position.

FLEXOR DIGITORUM BREVIS

Attachments: Attaches to the medial tubercle of the calcaneus and plantar fascia and extends, by four tendons, to the middle phalanges of the lateral four toes.
Nerve supply: Medial plantar S2, 3.

Surface markings: The muscle belly can be palpated on the plantar surface of the foot just distal to the calcaneus.

Actions
Rostral end fixed: It assists with flexion of the proximal interphalangeal joints of the lateral four toes.
Caudal end fixed: It assists in stabilizing the forefoot during weight bearing in the 'en pointe' position.

FLEXOR DIGITORUM ACCESSORIUS

Attachments: Extends from the medial and lateral borders of the inferior surface of the calcaneus to the tendons of flexor digitorum longus.
Nerve supply: Lateral plantar S2, 3.

Surface markings: Too deep to be palpated.

Actions
Rostral end fixed: It assists in flexing the second, third, fourth and fifth toes.
Caudal end fixed: It assists in stabilizing the forefoot during weight bearing in the 'en pointe' position.

For flexor digitorum longus, brevis and accessorius

Maximum extensibility: With the ankle joint dorsiflexed and the fore foot inverted, the lateral four toes are extended.

TESTING POSITION
Grades 5, 4 and 3 (Fig. 4.40): In supine lying or long sitting, with the foot supported in a plantargrade position, the interphalangeal joints of the lateral four toes are flexed. Resistance is applied to the plantar surface of the distal phalanges of the lateral four toes in the direction of extension.
Grades 2, 1 and 0 (Same position as for Fig. 4.36, p. **104**): In side lying, with the leg to be tested uppermost and supported, the interphalangeal joints of the lateral four toes are flexed.

Comments: The testing position for Grades 5, 4 and 3 is not against gravity, but this position is considered more practical as the patient can observe the movement. Vision is frequently important in assisting muscle action in the presence of weakness.

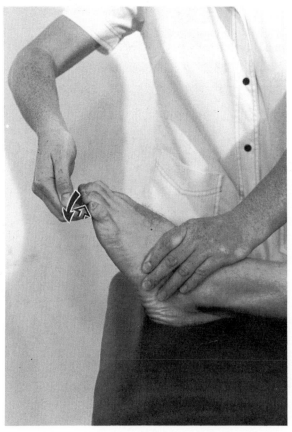

Fig. 4.40

ABDUCTOR HALLUCIS

Attachments: Extends from the medial process of the calcaneal tuberosity, the flexor retinaculum and the plantar aponeurosis to the medial side of the base of the proximal phalanx of the great toe. *Nerve supply*: Medial plantar S2, 3.

Surface markings: It can be palpated on the medial surface of the foot just proximal to the metatarsophalangeal joint of the great toe.

Actions

Rostral end fixed: It abducts the great toe and also assists in flexion of the metatarsophalangeal joint and adduction of the forefoot.

Caudal end fixed: It assists in stabilizing the great toe during weight bearing in the 'en pointe' position.

Maximum extensibility: With the forefoot pronated and the first metatarsophalangeal joint extended, the great toe is adducted.

TESTING POSITION

Grades 5, 4 and 3 (Not illustrated): In side lying, with the leg to be tested undermost·and supported, the metatarsophalangeal joint of the great toe is abducted. Resistance is applied to the medial surface of the distal phalanx of the great toe in the direction of adduction. *Grades 2, 1 and 0* (Fig. 4.41): In supine lying or long sitting, with the foot supported in a plantargrade position, the great toe is abducted at the metatarsophalangeal joint.

Comments: This is a very difficult isolated movement. Many people with normal muscle power will find this muscle action hard to demonstrate.

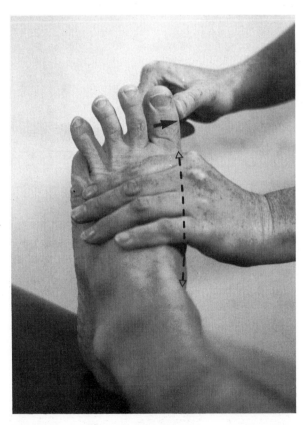

Fig. 4.41

Lower extremity

Muscles tested at the joints of the toes

ABDUCTOR DIGITI MINIMI

Attachments: Extends from the medial and lateral calcaneal tubercles to the lateral side of the proximal phalanx of the little toe.
Nerve supply: Lateral plantar S2, 3.

Surface markings: It can be palpated between its attachments over the lateral surface of the sole of the foot.

Actions
Rostral end fixed: It abducts the little toe.
Caudal end fixed: It assists in stabilizing the forefoot during weight bearing in the 'en pointe' position.

Maximum extensibility: With the foot in a plantargrade position, the little toe is adducted.

DORSAL INTEROSSEI

Attachments: All four muscles attach proximally by two heads to the shafts of the metatarsal bones. They attach caudally to the bases of the proximal phalanges and dorsal extension expansions of the second, third and fourth toes.
Nerve supply: Lateral plantar S2, 3.

Surface markings: The muscles are too deep to be palpated.

Actions
Rostral end fixed: They abduct the second, third and fourth toes and assist in flexion of the metatarsophalangeal joints of the same toes. They may assist in extension of the interphalangeal joints.
Caudal end fixed: They assist in stabilizing the toes during weight bearing in the 'en pointe' position.

Maximum extensibility: With the foot in a plantargrade position and the metatarsophalangeal joints extended, the second, third and fourth toes are adducted.

For abductor digiti minimi and dorsal interossei

TESTING POSITION
Grades 5, 4 and 3 (Fig. 4.42): In supine lying or long sitting, the metatarsophalangeal joints of the second to fifth toes are abducted. No resistance is applied.
Grades 2, 1 and 0 (Not illustrated): In supine lying or long sitting, the patient attempts to abduct the metatarsophalangeal joints of the second to fifth toes.

Comments: It is very difficult to isolate this movement. Many people with normal muscle power will find the movement hard to demonstrate.

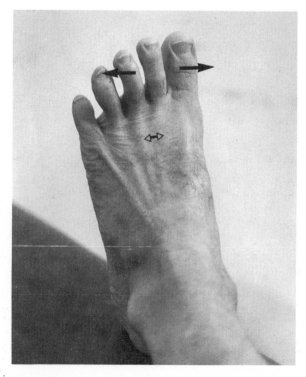

Fig. 4.42

110

ADDUCTOR HALLUCIS

Attachments: The oblique head extends from the lateral cuneiform bone and the bases of the third and fourth metatarsal bones to the lateral side of the base of the proximal phalanx of the great toe. The transverse head extends from the capsules of the lateral three metatarsophalangeal joints to the lateral side of the base of the proximal phalanx of the great toe.

Nerve supply: Lateral plantar S2, 3.

Surface markings: Too deep to be palpated.

Actions

Rostral end fixed: It adducts the great toe.
Caudal end fixed: It assists in stabilizing the great toe during weight bearing in the 'en pointe' position.

Maximum extensibility: With the foot in a plantargrade position, the great toe is abducted.

TESTING POSITION

Grades, 5, 4 and 3 (Fig. 4.43): With the lower leg supported, the great toe is adducted towards the second toe. Resistance is applied to the lateral surface of the proximal phalanx of the same toe in the direction of abduction.

Grades 2, 1 and 0 (Not illustrated): With the lower leg supported, the great toe is adducted towards the second toe.

Comments: This is a very difficult movement for the majority of subjects to isolate.

PLANTAR INTEROSSEI

Attachments: These muscles extend from the medial side of the third, fourth and fifth metatarsal bones to the medial side of the proximal phalanx of the same toes.

Nerve supply: Lateral plantar S2, 3.

Surface markings: Too deep to be palpated.

Actions

Rostral end fixed: They adduct the lateral three toes and assist in flexion of the metatarsophalangeal joints of the same toes.

Caudal end fixed: They aid in forefoot stability, as when the ballet dancer is 'en pointe'.

Maximum extensibility: All toes should be fully abducted while the metatarsophalangeal joints are flexed. It should be possible to insert a finger between each toe.

TESTING POSITION

Grades 5, 4 and 3 (Same position as Fig. 4.44): With the lower leg supported, the lateral three toes are adducted. Resistance is applied to the medial surface of the lateral three toes in the direction of abduction.

Grades 2, 1 and 0 (Fig. 4.44): With the lower leg supported, the lateral three toes are adducted.

Comments: As this is a difficult movement to isolate, it can be tested at the same time as the lumbricals and adductor hallucis.

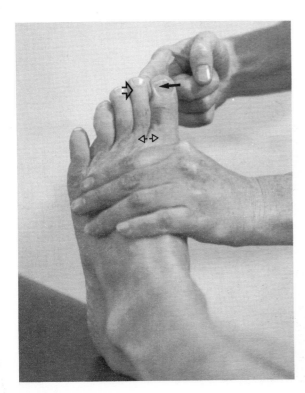

Fig. 4.43

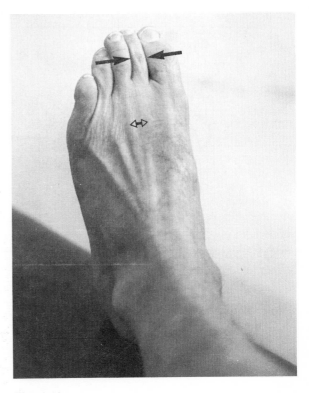

Fig. 4.44

Lower extremity

Muscles tested at the joints of the toes

LUMBRICALS

Attachments: The first extends from the medial side of the tendon of flexor digitorum longus to the second toe; the second, third and fourth from the adjacent sides of all the tendons of flexor digitorum longus. All attach to the bases of the proximal phalanxes of the lateral four toes.

Nerve supply: First lumbrical, medial plantar S2, 3; second, third and fourth, lateral plantar S2, 3.

Surface markings: Too deep to be palpated.

Actions

Rostral end fixed: They flex the metatarsophalangeal joints of the lateral four toes and assist in the extension of the interphalangeal joints of the same toes.

Caudal end fixed: For the ballet dancer, they assist in providing stability for the foot in the 'en pointe' position.

Maximum extensibility: The metatarsophalangeal joints of the lateral four toes are extended and abducted.

TESTING POSITION

Grades 5, 4 and 3 (Fig. 4.45): With the leg supported, the metatarsophalangeal joints of the lateral four toes are flexed. Resistance is applied to the plantar surface of the proximal phalanx of the same four toes.

Grades 2, 1 and 0 (Not illustrated): With the leg supported, the metatarsophalangeal joints are flexed.

Comments: During testing, the interphalangeal joints of the toes must remain in the neutral position, or the lumbrical action will not be seen. Figure 4.46 shows an alternative weight bearing position for testing in which some subjects might find the movement easier to produce.

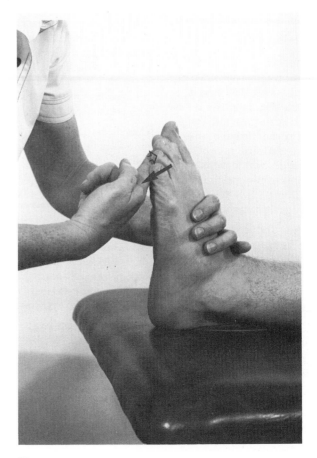

Fig. 4.45

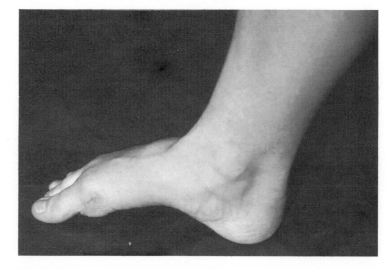

Fig. 4.46

MUSCLES WHICH ABDUCT THE
HIP JOINT
  Gluteus medius
    Gluteus minimus

MUSCLES WHICH EXTEND THE
KNEE JOINT
  Quadriceps femoris
    Vastus lateralis
    Vastus medialis
    Vastus intermedius
    Rectus femoris

MUSCLES WHICH PLANTAR FLEX
THE ANKLE JOINT
  Gastrocnemius
  Soleus

Lower extremity

Weight bearing tests of the lower extremity

GLUTEUS MEDIUS (See p. **87** for attachments, nerve supply, surface marking, actions, maximum extensibility)

FUNCTIONAL TESTING POSITION (Fig. 4.47): In normal standing, weight is transferred to single leg support. The iliac spines should remain level during the transfer, and the hip joint of the unsupported leg in abduction. Where weakness of this muscle is present on the side of the standing leg, the pelvis will drop towards the unsupported side and the freely hanging thigh will move into a position of adduction (Fig. 4.48); this is a positive Trendelenberg sign for the left gluteus medius.

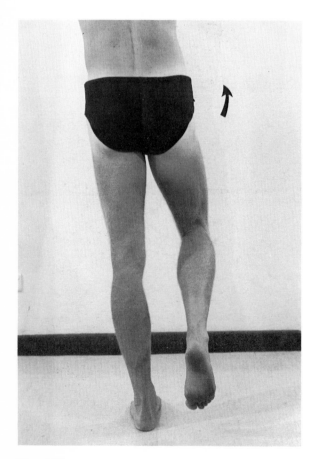

Fig. 4.47

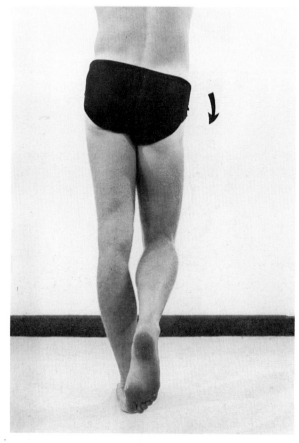

Fig. 4.48

Weight bearing tests of the lower extremity

QUADRICEPS FEMORIS (See p. **94** for attachments, nerve supply, surface marking, actions and maximum extensibility)

FUNCTIONAL TESTING POSITION (Fig. 4.49): From normal standing, the patient squats and returns to the upright position. Where there is weakness, the subject will have difficulty assuming the squatting position as well as being unable to extend the knees against the body weight.

GASTROCNEMIUS (See p. **98** for attachments, nerve supply, surface marking, actions and maximum extensibility)

FUNCTIONAL TESTING POSITION (Fig. 4.50): From normal standing, weight is transferred to single leg support. With the knee extended, the ankle is plantar flexed so as to take weight on the ball of the foot. Where weakness of the muscle is present, the subject will be unable to plantar flex the foot against the body weight without assistance.

SOLEUS (See p. **97** for attachments, nerve supply, surface marking, actions and maximum extensibility)

FUNCTIONAL TESTING POSITION (Not illustrated): From normal standing, weight is transferred to single leg support. With the knee flexed, the ankle is plantar flexed so as to take weight on the ball of the foot. Where weakness of the muscle is present, the subject will be unable to plantar flex the foot against the body weight without assistance.

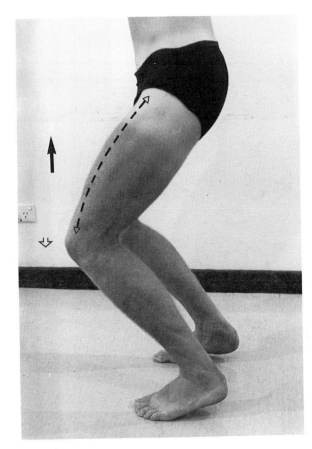

Fig. 4.49

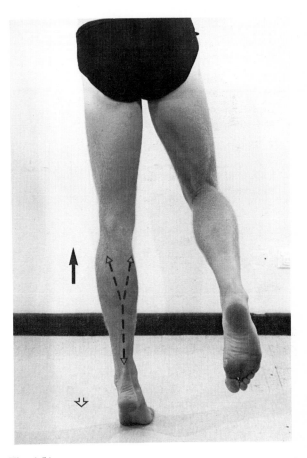

Fig. 4.50

5. Evaluating the muscles of the neck and trunk

Muscles which move the joints of the cervical spine (Fig. 5.1)

EXTENSION OF THE HEAD AND
NECK
   Superficial muscles:
      Splenius — capitis and cervicis
        Trapezius — upper fibres (Tuf)
   Intermediate muscles:
      Spinalis — capitis and cervicis
      Semispinalis — capitis and cervicis
      **Longissimus — capitis and
        cervicis**
      Iliocostalis — cervicis
   Deep muscles:
      Obliquus capitis inferior
      Obliquus capitis superior
      Rectus capitis major
      Rectus capitis minor
      Interspinales
      Intertransversarii
      Rotatores

FLEXION OF THE HEAD AND
NECK
   Longus colli
   Longus capitis
   Sternocleidomastoid (SM)
   Scalenus anterior
   Scalenus medius
   Scalenus posterior
   Rectus capitis lateralis
   Rectus capitis anterior

LATERAL FLEXION OF THE
HEAD AND NECK
   **Unilateral action of the neck flexors
   and extensors, and levator scapulae.**

ROTATION OF THE HEAD AND
NECK
   **Unilateral action of the neck flexors
   and extensors.**

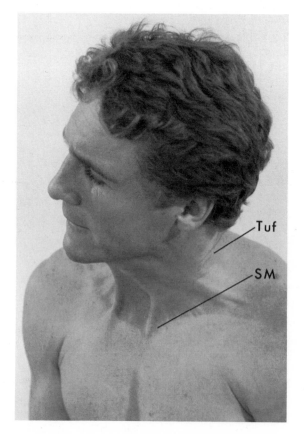

Fig. 5.1

Muscles which move the joints of the cervical spine

SPLENIUS CAPITIS

Attachments: Extends from the spinous processes of the fourth, fifth, sixth and seventh cervical vertebrae and also from those of the first, second and third thoracic vertebrae to the mastoid process and lateral half of the superior nuchal line of the occipital bone.
Nerve supply: Ventral rami of C3, 4, 5.

SPLENIUS CERVICIS

Attachments: Extends from the spinous processes of the third, fourth and fifth cervical vertebrae to the posterior tubercles of the first and second cervical vertebrae.
Nerve supply: Ventral rami of C6, 7, 8.

SPINALIS CAPITIS, SEMISPINALIS CAPITIS

Attachments: They extend from the transverse processes of the lower cervical and upper thoracic vertebrae to the occiput, and blend with each other.
Nerve supply: Dorsal rami of the cervical spinal nerves.

SPINALIS CERVICIS, SEMISPINALIS CERVICIS

Attachments: They extend from the transverse processes of the thoracic and lower cervical vertebrae to the spinous process of the axis, and to the spinous processes of the middle cervical vertebrae.
Nerve supply: Dorsal rami of the cervical spinal nerves.

LONGISSIMUS CAPITIS

Attachments: Extends from the transverse processes of the upper thoracic and lower cervical vertebrae to the mastoid process.
Nerve supply: Dorsal rami of the cervical spinal nerves.

LONGISSIMUS CERVICIS

Attachments: Extends from the transverse processes of the upper six thoracic vertebrae to the transverse processes of the second to sixth cervical vertebrae.
Nerve supply: Dorsal rami of the lower cervical and upper thoracic spinal nerves.

ILIOCOSTALIS CERVICIS

Attachments: Extends from the angles of the upper six ribs to the posterior tubercles of the transverse processes of third to the sixth cervical vertebrae.
Nerve supply: Dorsal rami of the cervical spinal nerves.

OBLIQUUS CAPITIS INFERIOR

Attachments: Extends from the apex of the spinous process of the axis to the inferior and posterior part of the transverse process of the atlas.
Nerve supply: Dorsal ramus of C1.

OBLIQUUS CAPITIS SUPERIOR

Attachments: Extends from the superior surface of the transverse process of the atlas to an attachment in between the superior and inferior nuchal lines of the occipital bone.
Nerve supply: Dorsal ramus of C1.

RECTUS CAPITIS POSTERIOR MAJOR

Attachments: Extends from the spinous process of the axis to the lateral part of the inferior nuchal line of the occipital bone.
Nerve supply: Dorsal ramus of C1.

RECTUS CAPITIS POSTERIOR MINOR

Attachments: Extends from the spinous tubercle of the posterior arch of the atlas to the medial part of the inferior nuchal line of the occipital bone.
Nerve supply: Dorsal ramus of C1.

INTERSPINALES

Attachments: These small muscles extend in pairs between the spinous processes of the vertebrae. (There are six cervical pairs, two or three thoracic pairs and four lumbar pairs.)
Nerve supply: Dorsal rami of the levels of attachment.

INTERTRANSVERSARII CERVICAL

Attachments: These small muscles extend between the transverse processes of the cervical vertebrae. They are present at all levels of the vertebral column.
Nerve supply: Dorsal rami of the levels of attachment.

ROTATORES CERVICAL

Attachments: Extend from the transverse processes of vertebrae and each inserts into the lamina of the vertebra above. They are present at all levels of the vertebral column.

(contd)

Neck and trunk

Muscles which move the joints of the cervical spine

For all neck extensors

Surface markings: Trapezius covers all the neck extensors, therefore they are too deep to be palpated.

Actions
Rostral end fixed: They assist in stabilizing of the neck in such activities as head-stands or somersaults.
Caudal end fixed: Bilaterally, they will extend the neck. Unilaterally, splenius capitis and cervicis, longissimus capitis and obliquus capitis inferior rotate the head to the same side. Splenius capitis and cervicis, iliocostalis cervicis, longissimus cervicis and obliquus capitis superior, laterally flex the neck to the same side. Semispinalis capitis and cervicis rotate the head to the opposite side. The interspinales and intertransversarii act as stabilizers of the vertebral column.

Maximum extensibility: In sitting or supine lying, the head is flexed, rotated and side flexed. Each of these movements should be performed both separately and in combinations.

TESTING POSITION
Grades 5, 4 and 3 (Fig. 5.2): In prone lying, with the head flexed over the end of the supporting surface, the neck is extended. Resistance is applied to the occiput in the direction of flexion.
Grades 2, 1 and 0 (Fig. 5.3): In side lying, with the head supported, the neck is extended.

Comments: Movements of the head and neck are carried out by an extremely complex group of muscles. It is impossible to isolate individual muscle actions when testing in this manner.

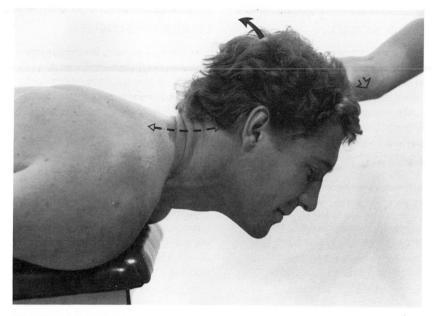

Fig. 5.2

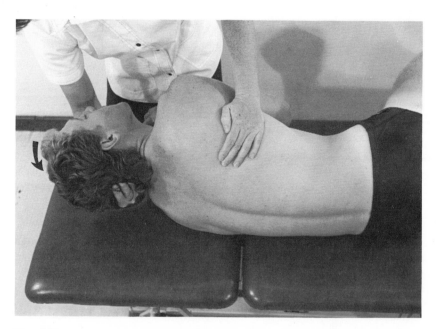

Fig. 5.3

LONGUS COLLI

Attachments: The superior oblique portion extends from anterior tubercles of the transverse processes of the third, fourth and fifth cervical vertebrae to the tubercle on the anterior arch of the atlas. The inferior oblique portion extends from the anterior surface of the bodies of the first, second and third thoracic vertebrae to the anterior tubercles of the transverse processes of the fifth and sixth cervical vertebrae. The vertical portion extends from the anterior surface of the bodies of the first, second and third thoracic vertebrae and the fifth, sixth and seventh cervical vertebrae to the anterior surface of the bodies of the second, third and fourth cervical vertebrae.
Nerve supply: Ventral rami of C2–6.

LONGUS CAPITIS

Attachments: Extends from the anterior tubercles of the transverse processes of the third, fourth, fifth and sixth cervical vertebrae to the inferior surface of the basilar process of the occipital bone.
Nerve supply: Ventral rami C1–3.

RECTUS CAPITIS ANTERIOR

Attachments: Extends from the root of the transverse process and anterior surface of the atlas to the inferior surface of the basilar tubercle of the occipital bone.
Nerve supply: Ventral rami of C1, 2.

STERNOCLEIDOMASTOID

Attachments: The sternal head extends from the cranial surface of the manubrium sterni. The clavicular head extends from the sternal third of the clavicle. Both heads attach to the lateral surface of the mastoid process and the lateral half of the nuchal line of the occipital bone.
Nerve supply: Spinal accessory (XIth cranial).

(contd)

Neck and trunk

Muscles which move the joints of the cervical spine

For all neck flexors

Surface markings: The longus colli and capitis and rectus capitis anterior are too deep to be palpated. The sternocleidomastoid is palpated easily between its attachments.

Actions

Rostral end fixed: These muscles assist in stabilizing the neck during such activities as heads stands and somersaults.

Caudal end fixed: Bilaterally, these muscles flex the head. Unilaterally, longus colli and sternocleidomastoid side flex the head to the same side and rotate it to the opposite side.

Maximum extensibility: In sitting or prone lying, the head is extended, laterally side flexed to the opposite side, and rotated towards the same side.

TESTING POSITION

Grades 5, 4 and 3 (Fig. 5.4): In supine lying, the chin is tucked in and the head flexed towards the chest. Resistance is applied to the chin in the direction of extension.

Grades 2, 1 and 0 (Same position as Fig. 5.3, p. **118**): In side lying, with the head supported, the chin is tucked in and the head flexed towards the chest.

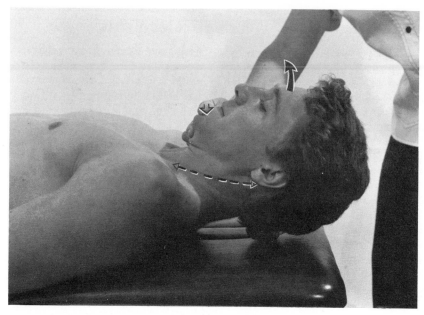

Fig. 5.4

Test position for left sternocleidomastoid

Grades 5, 4 and 3 (Fig. 5.5): In supine lying, the head is rotated to the right and side flexed to the left. Resistance is applied to the chin in the direction of left head rotation and to the left side of the head in the direction of right side flexion.

Grades 2, 1 and 0 (Not illustrated): The rotary component is tested in sitting. The head is rotated towards the right. The lateral side flexion component is tested in supine lying. The head is supported and is laterally side flexed towards the left.

Comments: It is important to test the sternocleidomastoids separately when asymmetry is present.

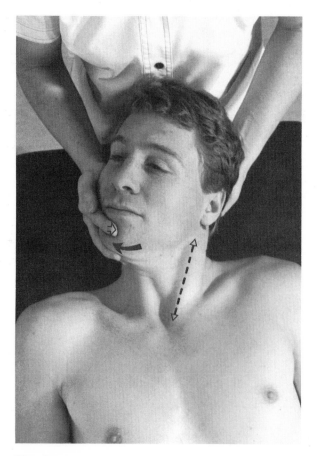

Fig. 5.5

Muscles which move the joints of the cervical spine

SCALENUS ANTERIOR

Attachments: Extends from the anterior tubercles of the transverse processes of the third, fourth, fifth and sixth cervical vertebrae to the scalene tubercle and cranial crest of the first rib.
Nerve supply: Ventral rami of C4, 5, 6.

SCALENUS MEDIUS

Attachments: Extends from the posterior tubercles of the transverse processes of the second to seventh cervical vertebrae to the cranial surface of the first rib, between the subclavian groove and the tubercle.
Nerve supply: Ventral rami of C3–8.

SCALENUS POSTERIOR

Attachments: Extends from the posterior tubercles of the transverse processes of the sixth and seventh (sometimes the fifth) cervical vertebrae to the outer surface of the second rib.
Nerve supply: Ventral rami of C6, 7, 8.

RECTUS CAPITIS LATERALIS

Attachments: Extends from the upper surface of the transverse process of the atlas to the inferior surface of the jugular process of the occipital bone.
Nerve supply: Ventral rami of C1, 2.

For all lateral neck muscles

Surface markings: Too deep to be palpated.

Actions
Rostral end fixed: Scalenus anterior and medius elevate the first rib. The scalenus posterior elevates the second rib. All three act as accessory respiratory muscles.
Caudal end fixed: Bilaterally, all these muscles, except rectus capitis lateralis, assist in neck flexion. Unilaterally, they side flex the head towards the same side and all, except rectus capitis lateralis, rotate the head towards the opposite side.

Maximum extensibility: In sitting or supine lying, the head is extended, laterally side flexed to the opposite side and rotated towards the same side.

TESTING POSITION
Grades 5, 4 and 3 (Fig. 5.6): In side lying, the head is side flexed towards the ceiling. Resistance is applied to the side of the head in the direction of side flexion towards the supporting surface.
Grades 2, 1 and 0 (Not illustrated): In supine lying, with the head supported, the head is side flexed.

Comments: The shoulder must be stabilized and should not be moved closer to the ear during testing.

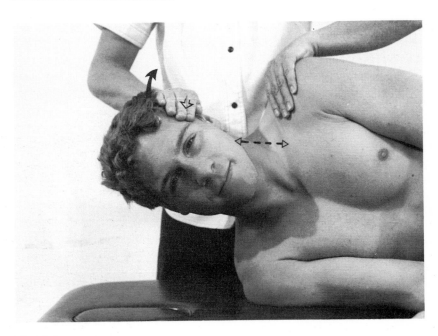

Fig. 5.6

Neck and trunk

Muscles which move the joints of the thoracic, sacral and lumber spine (Fig. 5.7)

EXTENSION OF THE TRUNK
   Superficial layer:
      Iliocostalis, thoracis and
        lumborum
      Longissimus, thoracis and
        lumborum
      Spinalis thoracis
   Intermediate layer:
      Semispinalis thoracis
      Multifidus
   Deep layer:
      Rotatores
      Interspinales
      Intertransversarii

FLEXION OF THE TRUNK
   Rectus abdominis (RA)
   Transversus abdominis
   Obliquus externus abdominis (EOA)
   Obliquus internus abdominis (IOA)

ROTATION OF THE TRUNK
   Obliquus externus abdominis
   Obliquus internus abdominis

ELEVATION OF THE PELVIS
   Quadratus lumborum

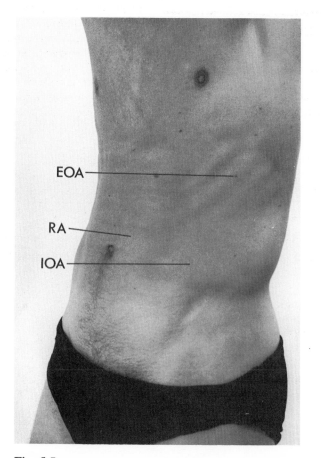

Fig. 5.7

Muscles which move the joints of the thoracic, sacral and lumbar spine

ILIOCOSTALIS THORACIS

Attachments: Extends from the dorsal surface of the erector spinae aponeurosis and the lower six ribs, just medial to the attachment of iliocostalis lumborum. The six fascicles sweep upwards to attach to the upper six ribs just beyond their angles.

ILIOCOSTALIS LUMBORUM

Attachments: Extends from the medial surface of the middle one-third of the iliac crest. The fibres are arranged in four fascicles directed towards a particular lumbar vertebra; the deepest to L4, the next to L3 and so on. The deep fascicles insert into the dorsal tip of the transverse processes of L1 to L4 and to the thoracolumbar fascia, while the superficial fascicles insert into the lower six ribs just below their angles.

LONGISSIMUS THORACIS

Attachments: Arises as eleven pairs of muscles on each side from the thoracic transverse processes and the adjacent part of the ribs. The short muscle bellies unite, producing long tendons extending down into the lumbar spine and attaching to the spinous processes of L2 to S3. The tendons from lower levels attach lowest and are lateral to tendons from higher levels.

LONGISSIMUS LUMBORUM

Attachments: Arises by five fascicles from the erector spinae aponeurosis, the middle one-third of the iliac crest and the lumbar intermuscular septum. They attach into the laminae of the accessory processes and the dorsal surface of the transverse processes of the five lumbar vertebrae. The fascicle of L5 is shortest, deepest and most medial.

SPINALIS THORACIS

Attachments: Extends from the spinous processes of the lower thoracic and upper lumbar vertebrae to the spinous processes of the upper sixth to the eighth thoracic vertebrae.

SEMISPINALIS THORACIS

Attachments: Extends from the transverse processes of the lower thoracic vertebrae to the spinous processes of the fifth to seventh cervical vertebrae.

MULTIFIDUS

Attachments: The deep layer of this muscle arises from the caudal end of the laminae of the lower two thoracic and all lumbar vertebrae and attaches to the mammillary processes and the capsule of the zygapophysial joints two levels below. The superficial fascicles arise from each of the spinous processes (T11–L5) and are anchored below to the mammillary processes, the iliac crest and the posterior surface of the sacrum. The most superficial fascicles (from higher levels) overlap those beneath.

ROTATORES

Attachments: Extend from the transverse processes of one vertebra to the lamina of the vertebrae above. Present at most of the spinal levels.

INTERSPINALES

Attachments: Attach in pairs to the spinous processes of adjoining vertebrae. Present at most of the spinal levels.

INTERTRANSVERSARII

Attachments: Attach to the transverse processes of adjoining vertebrae and are present at most of the spinal levels.

(contd)

Neck and trunk

Muscles which move the joints of the thoracic, sacral and lumber spine

For all trunk extensors

Nerve supply: Dorsal rami of the spinal nerves at the appropriate level. (Note: The intertransversarii are also supplied in part by the ventral rami.)

Surface markings: The iliocostalis and the longissimus thoracis can be palpated paravertebrally. All the other muscles in this group are too deep to be palpated.

Actions
Rostral end fixed: Bilaterally, these muscles stabilize and extend the thoracic and lumbar spine.
Caudal end fixed: Bilaterally all these muscles extend the thoracic and lumbar spine. Unilaterally, they act as lateral flexors of the trunk. The intermediate and deep layers assist in stabilizing each segment of the spinal column.

Maximum extensibility: With the hips flexed and knees extended, as in long sitting, the head is curled towards the knees.

TESTING POSITION
Grades 5, 4 and 3 (Fig. 5.8): In prone lying, with the legs stabilized and the arms by the side of the body, the trunk is extended. Resistance is applied between the scapulae in the direction of flexion.
Grades 2, 1 and 0 (Not illustrated): In side lying, with the legs flexed at the hip and knees to stabilize the pelvis, the trunk is extended. The friction offered by the couch in this position, however, may present excessive resistance and it may be better to leave the patient in prone and observe and palpate for any muscle action which can be produced.

Comments: Other muscle testing books use tests for Grade 5 back extensors which include longer lever arms. This makes the movement too difficult for many patients. The test illustrated in Figure 5.8 is consistent with the gravity plus resistance formula used throughout this text. To test very fit individuals a modification such as illustrated in Figure 5.9 may be necessary. For further information about the attachments of the trunk extensor muscles the reader is referred to Bogduk (1980), Bogduk & Twomey (1987) and Twomey & Taylor (1987).

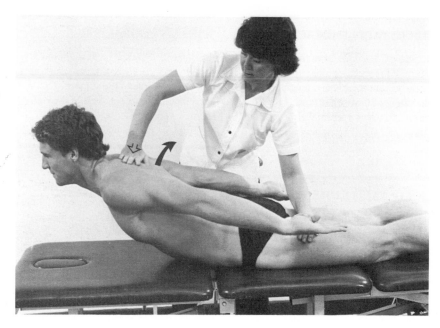

Fig. 5.8

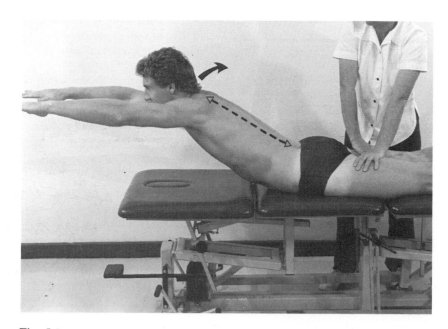

Fig. 5.9

RECTUS ABDOMINIS

Attachments: Extends from the crest of the pubic symphysis to the xiphoid process and the costal cartilages of the fifth, sixth and seventh ribs.
Nerve supply: Ventral rami of T5–12.

Surface markings: It is palpated easily between its attachments.

Actions
Rostral end fixed: It flexes the pelvis towards the thorax.
Caudal end fixed: It flexes the thorax towards the pelvis.

Maximum extensibility: In side lying, with the knees slightly flexed to stabilize the pelvis and the hips extended, the trunk is extended.

TESTING POSITION
Grades 5, 4 and 3 (Fig. 5.10): In supine lying, with the knees flexed about 60°, the thighs stabilized and the hands placed behind the head, the trunk is flexed towards the pelvis. Resistance is applied to the sternum in the direction of trunk extension.
Grades 2, 1 and 0 (Not illustrated): In side lying, with the legs flexed at the hip and knees to stabilize the pelvis, the trunk is flexed. The friction offered by the couch in this position, however, presents excessive resistance. In practice it is better to leave the patient in supine lying, with the arms by the side of the trunk, and palpate for any muscle action as the patient attempts to lift the head and shoulders from the supporting surface (Fig. 5.11).

Comments: By stabilizing the bent knees, the action of iliopsoas in assisting trunk flexion is enhanced. It is important to ascertain that the abdominal muscles are capable of performing the trunk flexion, by having the subject maintain both feet on the plinth without stabilization, before resistance is applied. All the abdominals will assist in this action.

TRANSVERSUS ABDOMINIS

Attachments: Extends from the inner surfaces of the seventh to twelfth costal cartilages, the lateral raphe of the

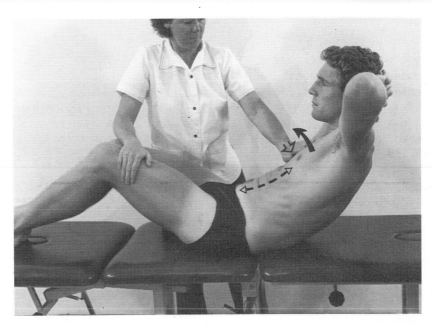

Fig. 5.10

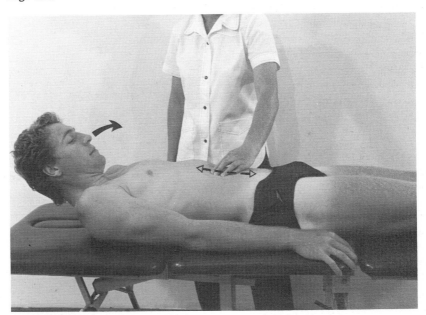

Fig. 5.11

thoracolumbar fascia, the anterior two-thirds of the iliac crest and the lateral third of the inguinal ligament, to the xiphoid cartilage, linea alba and via the falx inguinalis to the spine and crest of the pubis and iliopectineal line.
Nerve supply: Iliohypogastric, ilioinguinal, ventral rami of T7–12.

Surface markings: It is too deep to palpate.

Actions: It compresses the abdominal contents. It also stabilizes the linea alba, which provides better fixation for the other abdominal muscles.

Maximum extensibility: Not applicable.

TESTING POSITION
All grades: The muscle cannot be tested individually.

Comments: Weakness of this muscle results in bulging of the abdominal contents.

Neck and trunk

Muscles which move the joints of the thoracic, sacral and lumber spine

OBLIQUUS EXTERNUS ABDOMINIS

Attachments: Extends from the external surfaces of the fifth to twelfth ribs to the anterior half of the outer lip of the iliac crest, the inguinal ligament and the anterior layer of the rectus sheath.
Nerve supply: Ventral rami of T7–12.

Surface markings: It can be palpated near its attachments, just above the iliac crest or just under the ribs.

OBLIQUUS INTERNUS ABDOMINIS

Attachments: Extends from the anterior and middle one third of the iliac crest, the lateral two thirds of the inguinal ligament and the thoracolumbar fascia to the crest of the pubis, the medial part of the pectineal line, the linea alba (by means of an aponeurosis) and the inferior borders of tenth, eleventh and twelfth ribs.
Nerve supply: Ventral rami of T7–12, L1.

Surface markings: Too deep to be palpated.

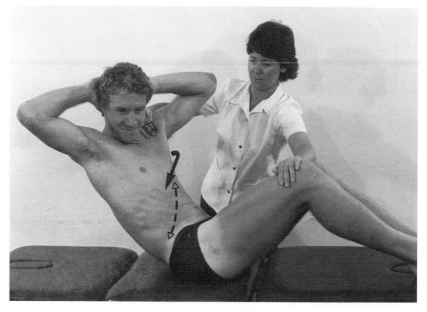

Fig. 5.12

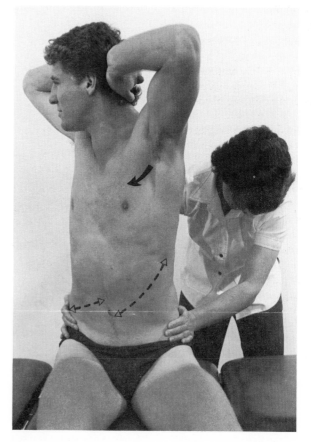

Fig. 5.13

126

For obliquus externus abdominis and obliquus internus abdominis

Actions

Rostral end fixed: Bilaterally, these muscles flex the pelvis towards the thorax and support and compress the abdominal contents. Unilaterally, they rotate the pelvis on the trunk. For example, the right internal oblique and left external oblique rotate the pelvis counter-clockwise.

Caudal end fixed: Bilaterally, these muscles flex the thorax towards the pelvis, support and compress the abdominal contents, and assist in respiration. Unilaterally, they rotate the thorax. For example, the right obliquus internus and left obliquus externus rotate the thorax clockwise.

Maximum extensibility: In sitting, with the legs over the edge of the supporting surface, the thorax is rotated.

In supine lying, with the arms abducted 90°, the flexed hips and knees are rolled to the left and right.

TESTING POSITION

Grades 5, 4 and 3 (Fig. 5.12): In supine lying, the knees flexed about 90°, the thighs stabilized, and the hands held behind the head, the left shoulder is flexed towards the right hip. Resistance is applied over the left pectoralis major in a counter-clockwise direction. Figure 5.12 shows the test in the middle to inner range of the movement.

Grades 2, 1 and 0 (Fig. 5.13): In sitting, with the legs over the edge of the supporting surface, both sides of the pelvis stabilized and the hands held behind the head, the trunk is rotated towards the right.

Comments: The action is reversed to test the right obliquus externus and the left obliquus internus.

Neck and trunk

Muscles which move the joints of the thoracic, sacral and lumber spine

QUADRATUS LUMBORUM

Attachments: Extends from the iliac crest and iliolumbar ligament to the medial portion of the lower border of the last rib and, by four tendinous slips, to the transverse processes of the upper four lumbar vertebrae.

Nerve supply: Lumbar plexus, T12–L4.

Surface markings: It can be palpated between its attachments lateral to the spinal muscles.

Actions

Rostral end fixed: It raises the pelvis towards the ribs.

Caudal end fixed: It side flexes the trunk, as in assuming a side sitting position.

Maximum extensibility: In side lying, with the muscle to be stretched undermost, the trunk is side flexed with weight being taken through the extended arm.

TESTING POSITION

Grades 5, 4 and 3 (Fig. 5.14): In standing, the pelvis is raised towards the shoulders, causing the pelvis to laterally tilt and the trunk to side flex slightly. Resistance is applied to the iliac crest in the direction of the floor.

Grades 2, 1 and 0 (Fig. 5.15): In prone lying, the pelvis is drawn towards the shoulder causing the pelvis to tilt laterally and the trunk to side flex slightly.

Comments: Grades 2, 1, and 0 can also be tested in supine lying. It is difficult, however, to palpate for muscle activity in this position.

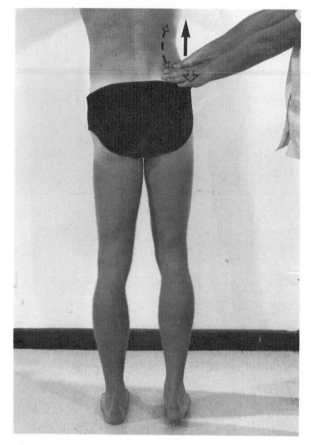

Fig. 5.14

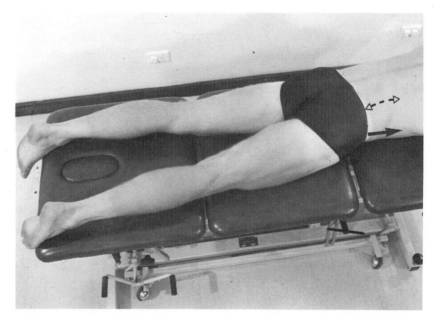

Fig. 5.15

128

Muscles of mastication (Fig. 6.1)

Masseter (M)
Temporalis (T)
Medial Pterygoid
Lateral Pterygoid

Muscles of facial expression (Fig. 6.2)

Epicranius — frontal bellies (Efb)
Corrugator supercilii (CS)
Levator palpebrae superioris
Orbicularis oculi (OOc)
Nasalis — alar portion
Nasalis — transverse portion (Nt)
Depressor septi
Procerus (Pr)
Levator anguli oris
Levator labii superioris (LLS)
Levator labii superioris alaeque nasi
Orbicularis oris (OOr)
Risorius (R)
Zygomaticus major   (Zmj
Zygomaticus minor   mn)
Mentalis (M)
Buccinator
Depressor anguli oris (DAO)
Platysma (P) (Fig. 6.1)
Depressor labii inferioris

Note: It is not practical to consider the muscles of mastication and facial expression as functioning with alternate ends fixed, since the variety of actions that can be produced by each is limited. Therefore one action only is considered. Grading these muscles on the six point scale presents problems as it is frequently not practical to perform the Grades 5, 4 and 3 tests against gravity, or to eliminate gravity from tests of Grades 2, 1 and 0. It is more sensible to grade the muscle as 2 if the action can be attempted, or 5 if it has normal function. Grade 0 is reserved for no function.

These muscles also differ from those of the trunk and extremities in that it is frequently difficult to isolate the unilateral action. For this reason many of the testing descriptions are based on bilateral muscle activity.

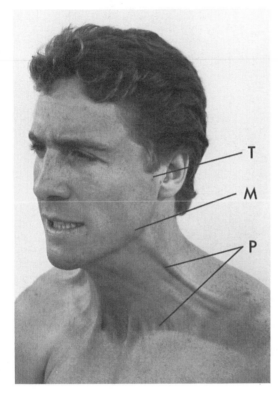

Fig. 6.1

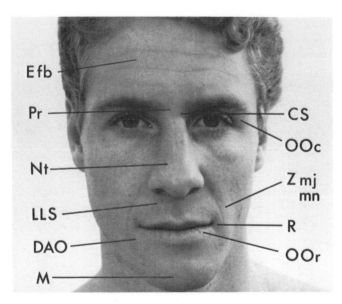

Fig. 6.2

Face and eyes

Muscles of facial expression

MASSETER

Attachments: Extends from the zygomatic process of the maxilla and the anterior two-thirds of the zygomatic arch to the lateral surface of the ramus and coronoid process of the mandible.
Nerve supply: Trigeminal (Vth cranial).

Surface markings: It can be palpated between its attachments when the jaws are clenched.

Actions: It closes and protrudes the jaw.

Maximum extensibility: The jaw is retracted and opened fully.

TEMPORALIS

Attachments: Extends from the temporal fossa and fascia of the cranium to the coronoid process and the anterior border of the ramus of the mandible.
Nerve supply: Trigeminal (Vth cranial).

Surface markings: It can be palpated just lateral to the orbital fossa.

Action: It closes and retracts the jaw.

Maximum extensibility: The jaw is protruded and opened fully.

MEDIAL PTERYGOID

Attachments: It extends from the medial surface of the lateral pterygoid plate and the tuberosity of the maxilla to the inferior and posterior part of the medial surface of the ramus and the angle of the foramen of the mandible.
Nerve supply: Trigeminal (Vth cranial).

Surface markings: It can be palpated on the medial side of the angle of the mandible.

Actions: It closes the jaw.

Maximum extensibility: The jaw is opened fully.

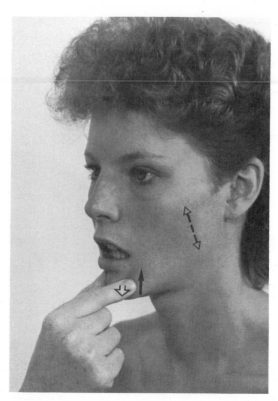

Fig. 6.3

For masseter, temporalis and medial pterygoid muscles

TESTING POSITION
Grade 5 (Fig. 6.3): The jaws are tightly closed, as in clenching the teeth. Resistance is applied to the chin in the direction of mouth opening.
Grade 2 (Not illustrated): The jaws are closed, as in clenching the teeth.

Comments: Gravity is not eliminated from the Grade 2 test unless the patient is in supine lying. These three muscles are tested together in the combined action of jaw closing. The protraction component of masseter and retraction component of temporalis may need to be evaluated separately if there is doubt about their function in individual patients.

LATERAL PTERYGOID

Attachments: The superior head attaches to the greater wing of the sphenoid and the inferior head to the lateral surface of the lateral pterygeroid plate. Both heads extend to the pterygoid fossa of the mandible and the anterior margin of the articular disc of the temporomandibular joint.
Nerve supply: Trigeminal (Vth cranial).

Surface markings: It is too deep to be palpated.

Actions: Bilaterally, it protrudes and opens the jaw. Unilaterally, it pulls the jaw horizontally to the side.

Maximum extensibility: The closed jaw is moved laterally as far as possible.

TESTING POSITION
Grade 5 (Fig. 6.4): The lower jaw is protruded. Resistance is applied to the chin in the direction of retraction. The unilateral action is tested by moving the open jaw laterally (not illustrated). Resistance is applied to the lateral surface of the chin in a horizontal direction.
Grade 2 (Not illustrated): The lower jaw is protruded. For the unilateral action, the open jaw is moved laterally.

Comments: This muscle is very important in chewing and should be examined carefully in patients with feeding difficulties.

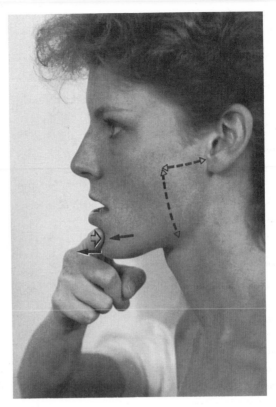

Fig. 6.4

Face and eyes

Muscles of facial expression

EPICRANIUS (FRONTAL BELLIES)

Attachments: Extends from the epicranial aponeurosis to blend medially with procerus and laterally with orbicularis oculi and corrugator.
Nerve supply: Facial (VIIth cranial).

Surface markings: It can be palpated on the forehead just above the eyebrows.

Actions: It draws the scalp backwards and the eyebrows upwards, wrinking the forehead.

Maximum extensibility: Not practicable.

TESTING POSITION
Grade 5 (Not illustrated): The eyebrows are lifted, causing horizontal wrinkles to appear in the forehead. Resistance is applied bilaterally over the eyebrows in the direction of the eyes.
Grade 2 (Fig. 6.5): The eyebrows are lifted, causing horizontal wrinkles to appear in the forehead.

Comments: Unilateral weakness is not observed often as the muscle is bilaterally innervated.

CORRUGATOR SUPERCILII

Attachments: Extends from the nasal part of the frontal bone to the skin of the eyebrow.
Nerve supply: Facial (VIIth cranial).

Surface markings: It can be palpated between the eyes.

Actions: It draws the skin of the forehead medially, causing vertical wrinkles to appear between the eyebrows.

Maximum extensibility: Not practicable.

TESTING POSITION
Grade 5 (Fig. 6.6): The eyebrows are drawn together, as in a frown, producing vertical wrinkles between the eyebrows.
Grade 2 (Not illustrated): The eyebrows are drawn together, as in a frown.

Comments: Applying resistance to this action is not practical.

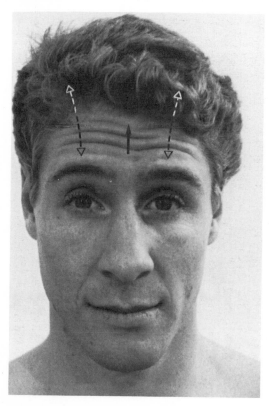

Fig. 6.5

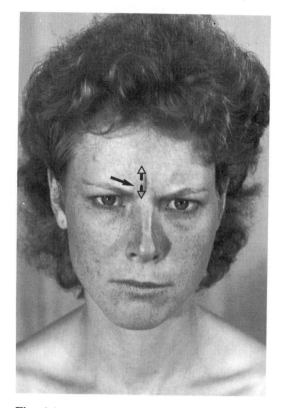

Fig. 6.6

LEVATOR PALPEBRAE SUPERIORIS

Attachments: Arises from the lesser wing of the sphenoid bone on its inferior surface and ends in a wide, divided aponeurosis in the skin of the eyelid.
Nerve supply: Oculomotor (IIIrd cranial).

Surface markings: It lies directly under the skin of the upper eyelid.

Actions: It opens the eyes widely.

Maximum extensibility: The upper eyelid is pulled down over the lower lid.

TESTING POSITION
Grade 5 (Fig. 6.7): The eyes are opened widely so that it is possible to see the white of the sclera above the iris.
Grade 2 (Not illustrated): The eyes are opened as widely as possible.

Comments: Most people can only perform this movement bilaterally.

ORBICULARIS OCULI

Attachments: Extends from the anterior surface of the medial palpebral ligament, the nasal part of the frontal bone and the frontal process of the maxilla to the medial palpebral ligament and medial orbital margin and the lateral palpebral raphe.
Nerve supply: Facial (VIIth cranial).

Surface markings: It can be palpated at the outer angle of the eye.

Actions: It closes the eye and draws the lid medially.

Maximum extensibility: The eyes are opened widely.

TESTING POSITION
Grade 5 (Fig. 6.8): The eyelid is closed tightly so that wrinkles form at the outer angle of the eye.
Grade 2 (Not illustrated): The eyelid is closed, but not so tightly as to form wrinkles at the outer angle of the eye.

Comments: As the muscle is innervated unilaterally each orbicularis oculi can be tested separately.

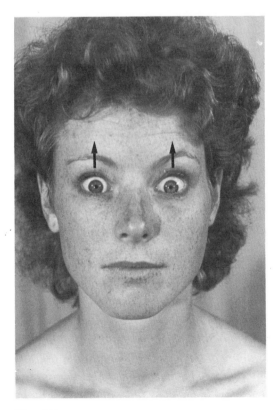

Fig. 6.7

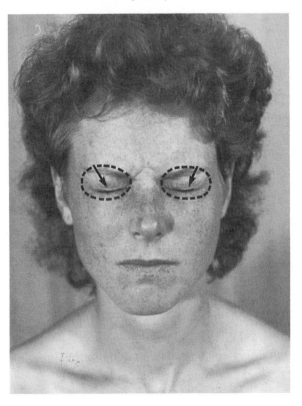

Fig. 6.8

Face and eyes

Muscles of facial expression

NASALIS, ALAR PORTION

Attachments: Extends from the greater alar cartilage to the crest of the nose.
Nerve supply: Facial (VIIth cranial).

Surface markings: It can be palpated as it moves the alar cartilage.

Actions: It widens the apertures of the nostrils.

Maximum extensibility: Not practicable.

TESTING POSITION
Grade 5 (Not illustrated): The nostrils are widened to make the apertures larger. Resistance is applied to the side of the nose in the direction of nasal closure.
Grade 2 (Fig. 6.9): The nostrils are widened to make the apertures larger.

Comments: Not all people are able to demonstrate the action of this muscle voluntarily.

NASALIS, TRANSVERSE PORTION

Attachments: Extends from above and lateral to the incisive fossa of the maxilla to the aponeurosis into which procerus attaches.
Nerve supply: Facial (VIIth cranial).

Surface markings: It can be palpated on the lateral aspect of the nose.

Actions: The point of the nose is drawn downwards such that the nostrils are narrowed.

DEPRESSOR SEPTI

Attachments: Extends from the incisive fossa of the maxilla and orbicularis oris to the septal cartilage of the nose.
Nerve supply: Facial (VIIth cranial).

Surface markings: It is too deep to be palpated.

Actions: It depresses the septum.

For transverse portion of nasalis and depressor septi

Maximum extensibility: Not applicable.

TESTING POSITION
Grade 5 (Fig. 6.10): The tip of the nose is drawn downwards and the nostrils are narrowed.
Grade 2 (Not illustrated): The tip of the nose is drawn downwards. The nostrils are narrowed marginally.

Comments: Not all people are able to perform this movement voluntarily. Applying resistance to this action is not practical.

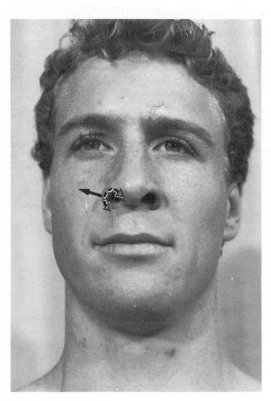

Fig. 6.9

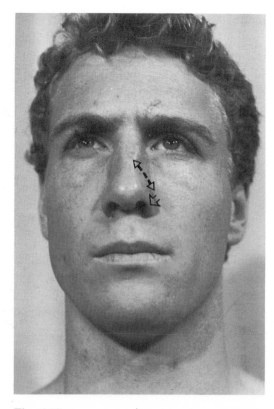

Fig. 6.10

PROCERUS

Attachments: Extends from the fascia over the lower part of the nose to the frontalis muscle and the skin between the eyes.
Nerve supply: Facial (VIIth cranial).

Surface markings: It can be palpated on the root of the nose just between the eyes.

Actions: It draws the skin at the root of the nose down forming a transverse wrinkle.

Maximum extensibility: Raising the eyebrows will elongate this muscle.

TESTING POSITION
Grade 5 (Fig. 6.11): The skin of the nose is pulled towards the eyes forming transverse wrinkles over the bridge of the nose.
Grade 2 (Not illustrated): The skin of the nose is pulled towards the eyes. Some transverse wrinkles may appear over the bridge of the nose.

Comments: Applying resistance to this action is not practical.

LEVATOR ANGULI ORIS

Attachments: Extends from the canine fossa of the maxilla to the angle of the mouth blending with orbicularis oris.
Nerve supply: Facial (VIIth cranial).

Surface markings: It can be palpated just below the lateral border of the nostril but may be difficult to distinguish from surrounding muscles.

Actions: It raises the angle of the mouth.

Maximum extensibility: The angles of the mouth are depressed.

TESTING POSITION
Grade 5 (Fig. 6.12): The upper lip is lifted, as in sneering. The wrinkle between the nose and the angle of the mouth should be accentuated and the gums visible.
Grade 2 (Not illustrated): The upper lip is lifted, as in sneering.

Comments: While it is easy to apply resistance to this muscle action, few people find it possible to perform the movement correctly against a resistance.

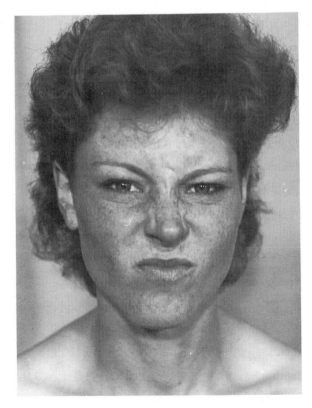

Fig. 6.11

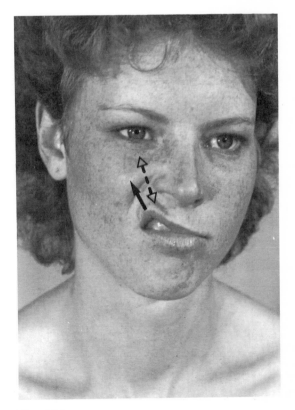

Fig. 6.12

Face and eyes

Muscles of facial expression

LEVATOR LABII SUPERIORIS

Attachments: Extends from the maxilla inferior to the infra-orbital foramen, to the orbicularis oris of the upper lip.
Nerve supply: Facial (VIIth cranial).

Surface markings: It can be palpated immediately beneath each nostril but is difficult to distinguish from surrounding muscles.

Actions: It elevates the upper lip.

Maximum extensibility: The upper lip is depressed.

LEVATOR LABII SUPERIORIS ALAEQUE NASI

Attachments: Extends from the upper part of the frontal process of the maxilla to the greater alar cartilage, skin of the nose and the lateral part of the orbicularis oris on the upper lip.
Nerve supply: Facial (VIIth cranial).

Surface markings: It can be palpated just lateral to the nostrils.

Actions: It raises the angle of the mouth and flares the nostril.

Maximum extensibility: The upper lip is depressed.

Levator labii superioris and levator superioris alaeque nasi

TESTING POSITION
Grade 5 (Fig. 6.13): The upper lip is raised and protruded to show the upper gums. The nostrils should flare.
Grade 2 (Not illustrated): The upper lip is raised and protruded.

Comments: It is not practical to apply resistance when testing this muscle.

ORBICULARIS ORIS

Attachments: It extends from adjacent muscles, mostly the buccinator, to the skin around the rim of the lip.
Nerve supply: Facial (VIIth cranial).

Surface markings: It can be palpated near the lips.

Actions: It closes, opens, purses and twists the mouth.

Maximum extensibility: The mouth is opened widely.

TESTING POSITION
Grade 5 (Not illustrated): The mouth is closed and the lips are protruded as in a whistle. Resistance is applied if the subject blows through a blocked straw.
Grade 2 (Fig. 6.14): The mouth is closed and the lips are protruded.

Comments: When weakness is present it is very difficult to control fluids in the mouth.

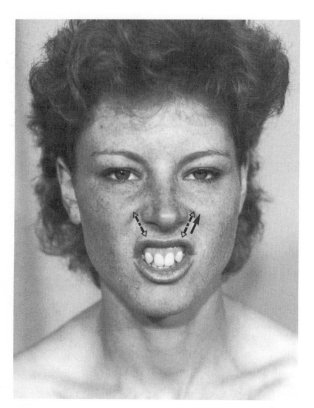

Fig. 6.13

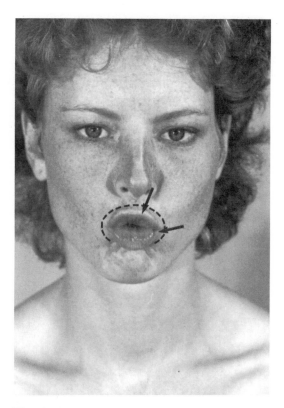

Fig. 6.14

RISORIUS

Attachments: Extends from the platysma, skin and fascia of masseter to the orbicularis oris and skin at the angle of the mouth.
Nerve supply: Facial (VIIth cranial).

Surface markings: It is too deep to be palpated.

Actions: It draws the corners of the mouth outwards, as in a grin.

Maximum extensibility: The lips are pursed.

TESTING POSITION
Grade 5 (Fig. 6.15): The corners of the mouth are drawn outwards. The wrinkle between the nose and the angle of the mouth is accentuated.
Grade 2 (Not illustrated): The corners of the mouth are drawn outwards.

Comments: Applying resistance to this action is not practical. The action of this muscle is closer to a grimace than a smile; it is important not to confuse it with the action of zygomaticus major and minor.

ZYGOMATICUS MINOR

Attachments: Extends from the zygomatic bone to the upper lip part of the orbicularis oris muscle.
Nerve supply: Facial (VIIth cranial).

ZYGOMATICUS MAJOR

Attachments: Extends from the zygomatic bone to the muscles at the angle of the mouth.
Nerve supply: Facial (VIIth cranial).

For zygomaticus minor and major

Surface markings: These muscles are too deep to be palpated.

Actions: They draw the upper lip diagonally upwards, as in a smile.

Maximum extensibility: The lips are pursed.

TESTING POSITION
Grade 5 (Fig. 6.16): The angle of the mouth is drawn upward and outward, as in a smile. The wrinkle between the nose and the angle of the mouth should be accentuated.
Grade 2 (Not illustrated): The angle of the mouth is drawn upward and outward, as in a smile.

Comments: Applying resistance to this action usually hinders the movement.

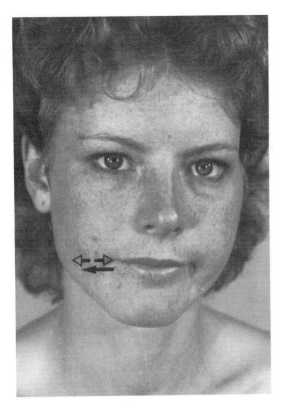

Fig. 6.15

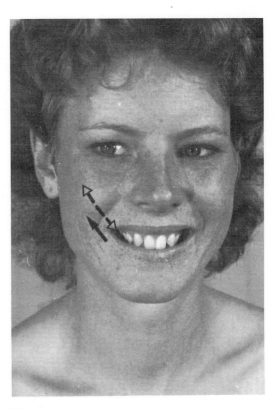

Fig. 6.16

Face and eyes

Muscles of facial expression

MENTALIS

Attachments: Extends from the incisor fossa of the mandible to the skin of the chin.

Nerve supply: Facial (VIIth cranial).

Surface markings: It can be palpated in the midline beneath the lower lip.

Actions: It raises and wrinkles the skin of the chin. This will cause the lower lip to protrude.

Maximum extensibility: Not practicable.

TESTING POSITION

Grade 5 (Fig. 6.17): The skin of the chin is raised and the lower lip protruded as if to pout.

Grade 2 (Not illustrated): The skin of the chin is raised and the lower lip may protrude slightly.

Comments: Applying resistance to this action is not practicable.

BUCCINATOR

Attachments: Extends from the outer surface of the alveolar processes of both the mandible and maxilla and the pterygomandibular ligament to the orbicularis oris at the corner of the mouth.

Nerve supply: Facial (VIIth cranial).

Surface markings: It can be palpated on the inner surface of the cheek.

Actions: It flattens the cheek against the teeth.

Maximum extensibility: The cheeks are blown outwards.

TESTING POSITION

Grade 5 (Fig. 6.18): Bilaterally, the cheeks are pulled in towards the lateral teeth. Resistance is applied in the opposite direction on the inner surface of the cheeks with the handle of a spoon.

Grade 2 (Not illustrated): The cheeks are pressed against the more lateral teeth.

Comments: The muscle is important in maintaining the food bolus between the teeth when chewing. It is likely to be well developed in those who play a wind or brass instrument.

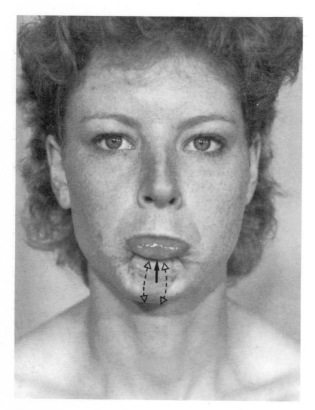

Fig. 6.17

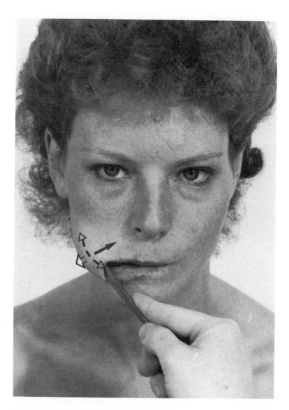

Fig. 6.18

DEPRESSOR ANGULI ORIS

Attachments: Extends from the oblique line of the mandible to the angle of the mouth blending with orbicularis oris.
Nerve supply: Facial (VIIth cranial).

Surface markings: It can be palpated immediately below the angle of the mouth but is difficult to distinguish from surrounding muscles.

Actions: It pulls down the corner of the mouth, as in an expression of sadness.

Maximum extensibility: Not applicable.

TESTING POSITION
Grade 5 (Fig. 6.19): The corner of the mouth is drawn downwards.
Grade 2 (Not illustrated): The corner of the mouth is drawn downwards slightly.

Comments: Most people find this muscle action difficult to perform unilaterally. Applying resistance to this action is not practical.

PLATYSMA

Attachments: Extends from the fascia that covers pectoralis major and deltoid to the inferior margin of the mandible and the skin of the lower part of the face and corners of the mouth.
Nerve supply: Facial (VIIth cranial).

Surface markings: It can be palpated and observed between its attachments.

Actions: It assists in depressing the jaw, lower lip and angle of the mouth. It also wrinkles the skin of the neck and upper part of the chest.

Maximum extensibility: The neck is extended with the mouth closed.

DEPRESSOR LABII INFERIORIS

Attachments: It extends from the oblique line of the mandible to the orbicularis oris and skin of the lower lip.
Nerve supply: Facial (VIIth cranial).

Surface markings: It can be palpated beneath the lower lip lateral to mentalis.

Actions: It depresses the lower lip.

Maximum extensibility: Not practicable.

For platysma and depressor labii inferioris

TESTING POSITION
Grade 5 (Fig. 6.20): The lower lip and angles of the mouth are drawn downwards and outwards and the skin over the neck is tensed.
Grade 2 (Not illustrated): The lower lip and angles of the mouth are drawn downwards and outwards.

Comments: Not all persons find it easy to isolate the action of platysma voluntarily. Applying resistance to this action is not practicable.

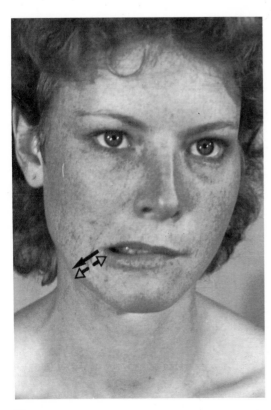

Fig. 6.19

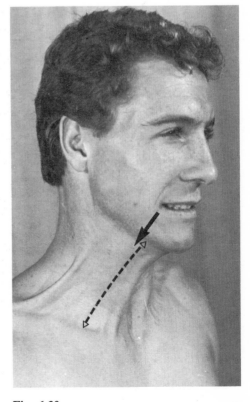

Fig. 6.20

Face and eyes

Muscles of the eye

Obliquus oculi superior
Obliquus oculi inferior
Rectus superior
Rectus inferior
Rectus medialis
Rectus lateralis

Note: It is not appropriate to consider surface markings and maximum extensibility of the muscles of the eye, therefore these headings have been omitted.

The strength of the muscles of the eyes cannot be assessed using the accepted procedures associated with muscle testing. The use of observation, however, can be of value in determining whether the patient can move the eyes or attempt to do so. To avoid the use of another scale, it is suggested a Grade 5 be allocated when the patient can move the eyes normally and Grade 2 when only an attempt can be made.

For a more detailed method of assessing the function of the eye muscles the reader is referred to Leigh & Zee (1983).

OBLIQUUS OCULI SUPERIOR

Attachments: Arises from the sphenoid bone above and medial to the optic canal and ends in a tendon which passes through a loop (the trochlea) and changes direction to insert in the sclera of the eye, behind the equator between the rectus superior and the rectus lateralis.
Nerve supply: Trochlear (IVth cranial).

Actions: It rotates the posterior surface of the eyeball upwards, moving the visual axis down towards the floor.

OBLIQUUS OCULI INFERIOR

Attachments: Arises from the orbital surface of the maxilla and is directed laterally, backwards and upwards to attach to the sclera of the eye behind the equator adjacent, but posterior, to the attachment of the superior oblique.
Nerve supply: Oculomotor (IIIrd cranial).

Actions: It rotates the posterior surface of the eyeball downwards, moving the visual axis upwards towards the ceiling.

RECTUS SUPERIOR

Attachments: Extends from the upper part of the common annular tendon to the sclera of the superior surface of the eye.
Nerve supply: Oculomotor (IIIrd cranial).

Actions: It elevates and medially rotates the eyeball.

RECTUS INFERIOR

Attachments: It extends from the lower portion of the common annular tendon to the sclera of the inferior surface of the eye.
Nerve supply: Oculomotor (IIIrd cranial).

Actions: It depresses and medially rotates the eyeball.

RECTUS MEDIALIS

Attachments: Extends from the adjacent aspects of both the upper and lower components of the common annular tendon to the sclera of the medial surface of the eye.
Nerve supply: Oculomotor (IIIrd cranial).

Actions: It medially rotates the eye.

RECTUS LATERALIS

Attachments: Extends from the adjacent parts of the upper and lower components of the common annular tendon to the sclera of the lateral surface of the eye.
Nerve supply: Abducent (VIth cranial).

Actions: It laterally rotates the eye.

(contd)

Face and eyes

Muscles of the eye

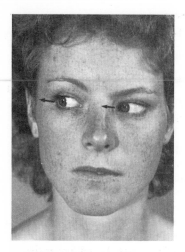

Fig. 6.21

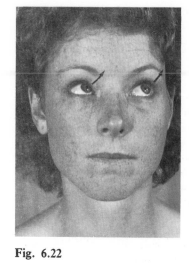

Fig. 6.22

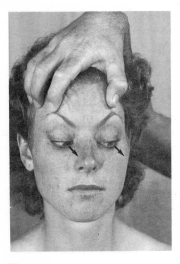

Fig. 6.23

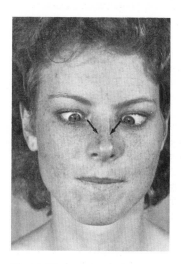

Fig. 6.24

For all the eye muscles

TESTING POSITION

Grade 5 : In half lying or sitting, the eyes follow a moving target to the right and left in the horizontal plane (Fig. 6.21 medial and lateral rectus), obliquely up to the ceiling (Fig. 6.22 superior rectus and inferior oblique), and obliquely down to the floor (Fig. 6.23 inferior rectus and superior oblique), in both the right and left diagonals.

Grade 2 : In half lying or sitting, the eyes attempt to follow a moving target to the right and left in the horizontal plane, (medial and lateral rectus), obliquely up to the ceiling, (superior rectus and inferior oblique), and obliquely down to the floor, (inferior rectus and superior oblique), in both the right and left diagonals.

Comments: All the muscles which attach to the sclera probably contribute in some degree to all movements of the eye. Therefore, it is difficult to isolate the specific contribution of each to the components of functional gaze. The ability to fix the gaze on the stimulus, maintain the stimulus in focus, and follow it rapidly and smoothly in the various directions should be noted. The subject's capacity for convergent movements should also be assessed by having the eyes follow a target which is approaching the nose (Fig. 6.24). Frequent blinking, shifting of the gaze or jerky movements are indicative of functional problems. Nystagmus or strabismus may also be observed.

SWALLOWING — STAGE 1. THE BOLUS MOVES BACKWARDS. AS THE HYOID BONE ASCENDS, THE FLOOR OF THE MOUTH TENSES.

Orbicularis oris
Buccinator
Superior longitudinal ⎤
Inferior longitudinal ⎟ **Intrinsic muscles of the tongue**
Transverse ⎟
Vertical ⎦
Genioglossus ⎤
Hyoglossus ⎟ **Extrinsic muscles of the tongue**
Chondroglossus ⎟
Styloglossus ⎦
Digastric ⎤
Stylohyoid ⎟ **Suprahyoids**
Mylohyoid ⎟
Geniohyoid ⎦

SWALLOWING — STAGE 2. THE PHARYNX IS ELEVATED AND CONSTRICTS OVER THE TOP OF THE BOLUS. THE THYROID CARTILAGE IS ELEVATED.

Tensor veli palatini
Levator veli palatini
Palatopharyngeus
Superior pharyngeal constrictor
Middle pharyngeal constrictor
Stylopharyngeus
Salpingopharyngeus
Aryepiglottic
Thyroarytenoid
Oblique arytenoid
Transverse arytenoid
Thyrohyoid

SWALLOWING — STAGE 3. THE BOLUS IS COMPRESSED ONWARDS INTO THE OESOPHAGUS.

Inferior pharyngeal constrictor
Sternohyoid ⎤
Sternothyroid ⎟ **Infrahyoids**
Thyrohyoid ⎟
Omohyoid ⎦

Note: It is not appropriate to consider surface markings and maximum extensibility of the muscles of swallowing, so these headings have been omitted.

The examination of swallowing should always be performed in the upright position. It is an activity which is normally assisted by gravity, therefore the upright position is optimal for testing. It also limits the possibility of aspiration during the examination. The protective reflexes, cough and gag, should be evaluated prior to the introduction of food to the mouth. In the absence of these reflexes, swallowing should be examined using alternative means of stimulation (for example, ice, pressure under the chin or stimulation of the taste buds).

The strength of the muscles involved in swallowing cannot be assessed using the accepted procedures associated with muscle testing. The use of observation and palpation, however, can be of value to determine whether the patient can swallow or attempt this action. To avoid the use of another scale it is suggested a Grade 5 be allocated when the patient can swallow and a Grade 2 when there is an attempt to swallow.

For more details of the sequence, and accurate methods of the assessment of swallowing, the reader is referred to Miller (1982).

Essential functions

Evaluating the muscles of swallowing – stage 1

ORBICULARIS ORIS (see p. 136)

BUCCINATOR (see p. 138)

SUPERIOR LONGITUDINAL OF THE TONGUE

Attachments: Intrinsic to the tongue.
Nerve supply: Hypoglossal (XIIth cranial).

Actions: It shortens and raises the sides and the tip of the tongue.

INFERIOR LONGITUDINAL OF THE TONGUE

Attachments: Intrinsic to the tongue.
Nerve supply: Hypoglossal (XIIth cranial).

Actions: It shortens and turns the tip of the tongue downwards.

TRANSVERSE OF THE TONGUE

Attachments: Intrinsic to the tongue.
Nerve supply: Hypoglossal (XIIth cranial).

Actions: It lengthens and narrows the tongue.

VERTICAL OF THE TONGUE

Attachments: Intrinsic to the tongue.
Nerve supply: Hypoglossal (XIIth cranial).

Actions: It flattens and broadens the tongue.

For superior and inferior longitudinal, transverse and vertical of the tongue

TESTING POSITION
Grade 5: In sitting, the tip of the tongue is placed onto the roof of the mouth (Fig. 7.1) and the floor of the mouth (not illustrated). It is also protruded (Fig. 7.2) and retracted (not illustrated).
Grade 2: In sitting, the patient attempts to place the tip of the tongue onto the roof of the mouth and the floor of the mouth. Protrusion and retraction of the tongue are also attempted.

Comments: These muscles, which make up the substance of the tongue, are responsible for altering its shape. It is difficult to apply resistance to them. When the tongue is protruded the assistance of genioglossus is required.

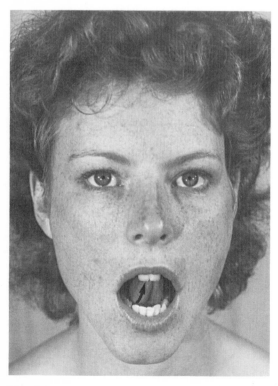

Fig. 7.1

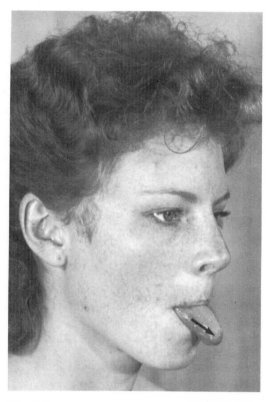

Fig. 7.2

GENIOGLOSSUS

Attachments: Extends from the genial spine of the mandible to the ventral surface of the tongue.
Nerve supply: Hypoglossal (XIIth cranial).

Actions: It protrudes and depresses the tongue.

TESTING POSITION
Grade 5 (See Fig. 7.2): In sitting, the tongue is protruded. Resistance is applied to the tip of the tongue, in a horizontal direction towards the mouth (the resistance is not shown).
Grade 2 (Not illustrated): In sitting or half lying, the patient attempts to protrude the tongue.

Comments: In some subjects, this muscle can depress the middle section of the tongue causing the edges to roll upwards, forming a deep groove (Fig. 7.3).

HYOGLOSSUS

Attachments: Extends from the body and greater horn of the hyoid bone to the lateral aspect of the tongue.
Nerve supply: Hypoglossal (XIIth cranial).

CHONDROGLOSSUS

Attachments: Extends from the lesser horn and body of the hyoid bone to the ventral surface of the tongue between the genioglossus and hyoglossus.
Nerve supply: Hypoglossal (XIIth cranial).

STYLOGLOSSUS

Attachments: It extends from the styloid process to blend with the side of the tongue and the hyoglossus.
Nerve supply: Hypoglossal (XIIth cranial).

For hypoglossus, chondroglossus and styloglossus

Actions: The hyoglossus and chondroglossus depress the tongue. The styloglossus retracts the tongue.

TESTING POSITION
Grade 5 (Not illustrated): In sitting, with the mouth open, the tongue is retracted and depressed against the floor of the mouth.
Grade 2 (Not illustrated): In sitting, with the mouth open, an attempt is made to retract and depress the tongue against the floor of the mouth.

Comments: This is a difficult action to perform and is best explained by having the patient say 'ah'. It is not possible to apply resistance to this action.

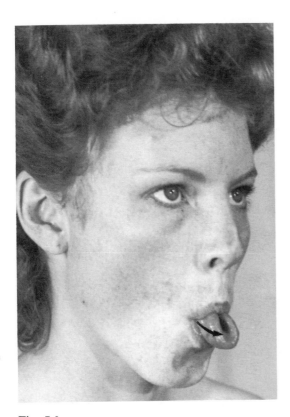

Fig. 7.3

Essential functions

PALATOGLOSSUS

Attachments: Extends from the oral surface of the soft palate to the lateral surface of the tongue.
Nerve supply: Vagus nerve (Xth cranial).

Actions: It raises the back of the tongue and narrows the fauces. This action closes off the oral cavity from the oropharynx.

TESTING POSITION
Grade 5 (Not illustrated): In sitting with the mouth open, the back of the tongue is elevated against the palate. Resistance is applied, with a spatula, to the back of the tongue in the direction of the floor of the mouth.
Grade 2 (Not illustrated): In sitting with the mouth open, the patient attempts to elevate the back of the tongue towards the palate.

Comments: When offering resistance to this movement care must be taken to ensure that the gag reflex is not elicited.

DIGASTRIC

Attachments: The posterior belly attaches from the digastic groove near the mastoid process and the anterior belly from the lower border of the mandible near the symphysis. Both bellies insert into an intermediate tendon which is held to the hyoid bone via a fibrous loop.
Nerve supply: Mylohyoid branch of the inferior alveolar (Vth cranial) and facial nerve (VIIth cranial).

STYLOHYOID

Attachments: Extends from the posterior surface of the styloid process of the temporal bone to the body of the hyoid bone.
Nerve supply: Facial nerve (VIIth cranial).

MYLOHYOID

Attachments: Extends from the mylohyoid line of the mandible to the body of the hyoid bone.
Nerve supply: Mylohyoid branch of the inferior alveolar nerve (Vth cranial).

GENIOHYOID

Attachments: Extends from the mental spine of the mandible to the anterior surface of the body of the hyoid bone.
Nerve supply: Cervical plexus, hypoglossal nerve (XIIth cranial).

For digastric, stylohyoid, mylohyoid and geniohyoid

Actions: These muscles raise the hyoid bone upwards during the first stage of swallowing. Except for the stylohyoid, they also depress the mandible.

TESTING POSITION
Grade 5: From the relaxed position in sitting (Fig. 7.4), the upward movement of the larynx is observed, as a bolus of food is swallowed (Fig. 7.5).
Grade 2: In sitting the upward movement of the larynx is observed and palpated as swallowing is attempted.

Comments: An alternative method of evaluating the suprahyoid group is to apply resistance to the secondary action of mandibular depression (Fig. 7.6). With the patient in sitting, the mouth is opened. Resistance is applied to the chin in the direction of mouth closing. As this movement is usually performed with the assistance of gravity, a gravity resisted position is inappropriate.

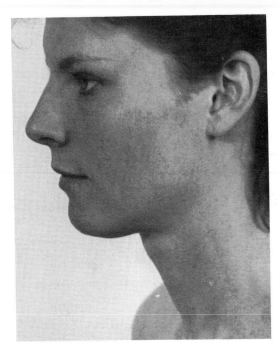

Fig. 7.4

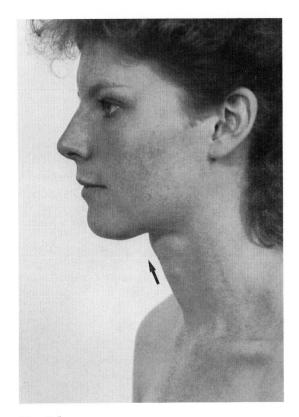

Fig. 7.5

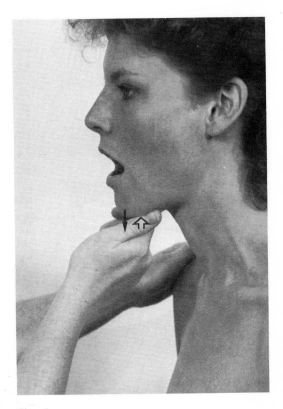

Fig. 7.6

Essential functions

Evaluating the muscles of swallowing – stage 2

TENSOR VELI PALATINI

Attachments: Extends from the scaphoid fossa and spine of the sphenoid bone, and the cartilage of the Eustachian tube to the posterior border of the hard palate and the aponeurosis of the soft palate.
Nerve supply: Mandibular branch of the trigeminal (Vth cranial).

Actions: It tenses the soft palate.

LEVATOR VELI PALATINI

Attachments: Extends from the apex of the petrous portion of the temporal bone and the cartilage of the eustachian tube to the aponeurosis of the soft palate.
Nerve supply: Pharyngeal plexus (IXth, Xth and XIth cranial).

Actions: It elevates the soft palate.

PALATOPHARYNGEUS

Attachments: Extends from the palate to the posterior border of the thyroid cartilage and the aponeurosis of the pharynx.
Nerve supply: Pharyngeal plexus (IXth, Xth and XIth cranial).

Actions: It narrows the fauces and it aids in the elevation of the pharynx during the second stage of swallowing.

SUPERIOR PHARYNGEAL CONSTRICTOR

Attachments: Extends from the pterygoid hamulus, pterygomandibular raphe, mylohyoid line of the mandible and the side of the tongue to the midline raphe of the posterior wall of the pharynx.
Nerve supply: Pharyngeal plexus (IXth, Xth and XIth cranial).

MIDDLE PHARYNGEAL CONSTRICTOR

Attachments: Extends from the stylohyoid ligament and the greater cornua of the hyoid bone to the middle of the posterior wall of the pharynx.
Nerve supply: Pharyngeal plexus (IXth, Xth and XIth cranial).

For the pharyngeal constrictors

Actions: These muscles sequentially narrow the nasopharynx, oropharynx and laryngopharynx during the second stage of swallowing.

STYLOPHARYNGEUS

Attachments: Extends from the styloid process to the posterior border of the thyroid cartilage and the posterolateral wall of the pharynx.
Nerve supply: Glossopharyngeal (IXth cranial).

Actions: It elevates the pharynx and larynx, during the second stage of swallowing.

SALPINGOPHARYNGEUS

Attachments: It extends from the auditory tube to blend with the palatopharyngens.
Nerve supply: Pharyngeal plexus (IXth, Xth and XIth cranial).

Actions: It elevates the pharynx during the second stage of swallowing.

ARYEPIGLOTTIC

Attachments: Extends from the apex of the arytenoid cartilage to the aryepiglottic fold.
Nerve supply: Vagus nerve (Xth cranial).

THYROARYTENOID

Attachments: Extends from the thyroid cartilage to the anterolateral surface of the arytenoid cartilage.
Nerve supply: Vagus nerve (Xth cranial).

OBLIQUE ARYTENOID

Attachments: Extends from the base of one arytenoid cartilage to the apex of the opposite arytenoid cartilage.
Nerve supply: Vagus nerve (Xth cranial).

TRANSVERSE ARYTENOID

Attachments: Extends from the posterior surface and lateral border of one arytenoid cartilage to the posterior surface and lateral border of the opposite arytenoid cartilage.
Nerve supply: Vagus nerve (Xth cranial).

For aryepiglotic, thyroarytenoid, oblique arytenoid and transverse arytenoid

Actions: These muscles assist in closing the glottis by adducting the arytenoid cartilages during the second stage of swallowing.

For all muscles of the second stage of swallowing

TESTING POSITION
Grade 5 (Fig. 7.4, p. **147**): In sitting, the upward movement of the larynx is observed as saliva or a bolus of food is swallowed.
Grade 2 (Fig. 7.4, p. **147**): In sitting, the upward movement of the larynx is observed and the thyroid cartilage is palpated as swallowing is attempted.

Comments: These muscles prevent the bolus passing upward to the nasopharynx or larynx. An indication of dysfunction is the tendency of the patient to cough or choke on attempting to swallow a bolus.

INFERIOR PHARYNGEAL CONSTRICTOR

Attachments: Extends from the thyroid and cricoid cartilages to the midline raphe at the posterior portion of the wall of the larynx.
Nerve supply: Pharyngeal plexus (IXth, Xth and XIth cranial).

Actions: As for pharyngeal constrictors (p. **148**)

STERNOHYOID

Attachments: Extends from the posterior surface of the medial end of the clavicle, the posterior sterno clavicular ligament and the superior and posterior part of the manubrium sterni to the lower border of the body of the hyoid bone.
Nerve supply: Upper cervical via the ansa cervicalis.

Actions: It depresses the hyoid bone in the third stage of swallowing.

STERNOTHYROID

Attachments: Extends from the posterior surface of the manubrium sterni, and the cartilage of the first rib to the oblique line of the lamina of the thyroid cartilage.
Nerve supply: Upper cervical via the ansa cervicalis.

Actions: It depresses the larynx during the third stage of swallowing.

THYROHYOID

Attachments: Extends from the oblique line of the lamina of the thyroid cartilage to the body of the hyoid bone.
Nerve supply: Upper cervical via the ansa cervicalis.

Actions: It approximates the hyoid bone to the larynx

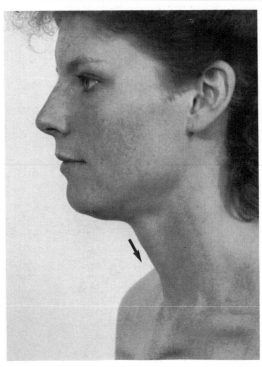

Fig. 7.7

OMOHYOID

Attachments: It extends, via an inferior belly, from the upper border of the scapula to the lower border of the hyoid bone, via a superior belly.
Nerve supply: Hypoglossal, (XII cranial).

Actions: It depresses the hyoid bone.

For infrahyoids

TESTING POSITION
Grade 5 (Fig. 7.7): In sitting, the downward movement of the larynx is observed as a bolus of food is swallowed.
Grade 2 (Not illustrated): In sitting, the downward movement of the larynx is observed and the thyroid cartilage palpated as swallowing is attempted.

Comments: The infrahyoid group stabilize the suprahyoid muscles in depression of the hyoid bone. They also serve to fix the hyoid bone to provide stability for the suprahyoid actions. In addition, the sternohyoid prevents the inward movement of the soft parts of the larynx during the inspiratory phase of breathing.

Essential functions

Muscles of respiration

Diaphragm
Intercostales — externi
             — interni
             — intimi
Subcostales
Transversus thoracis
Levatores costarum
Serratus posterior — inferior
                        — superior
   Scalenus anterior
   Scalenus medius
   Scalenus posterior
   Sternocleidomastoid

Note: It is not appropriate to consider surface markings or maximum extensibility in the primary respiratory muscles, so these headings have been omitted.

As with other muscles involved in involuntary functions, it is not possible to test the strength of these muscles using manual muscle testing procedures. Observation of movement of the thoracic wall does not provide information about the strength of the diaphragm or the intercostals.

For the assessment of the strength of these muscles, the reader is referred to Green & Moxham (1983).

DIAPHRAGM

Attachments: The sternal portion attaches to the back of the xiphoid process, the costal portion to the cartilages and adjacent parts of the lower six ribs (interdigitating with transverse abdominis), and the lumbar portion to the medial and lateral arcuate ligaments and the right and left crura, which in turn attach to the anterior surface of the lumbar vertebrae. All these portions insert into the central tendon of the diaphragm.
Nerve supply: Phrenic C3, 4, 5.

Actions: It descends when it contracts, increasing the volume and decreasing the pressure in the thoracic cavity and, conversely, decreasing the volume and increasing the pressure in the abdominal cavity.

INTERCOSTALES — EXTERNI

Attachments: Attach between adjacent borders of consecutive ribs, extending from the tubercles of the ribs almost to the cartilaginous attachments of the ribs at the sternum.
Nerve supply: Adjacent intercostal nerves

INTERCOSTALES — INTERNI

Attachments: Attach between adjacent borders of consecutive ribs, extending from the sternum to the posterior costal angles.
Nerve supply: Adjacent intercostal nerves

INTERCOSTALES — INTIMI

Attachments: Extend between the internal surfaces of adjoining ribs.
Nerve supply: Adjacent intercostal nerves.

All intercostales

Actions: Approximation of the ribs.

SUBCOSTALES

Attachments: Descend from the internal surface of the lower ribs near their angles to the second or third rib below.
Nerve supply: Adjacent intercostal nerves.

TRANSVERSUS THORACIS (STERNOCOSTALIS)

Attachments: It arises from the posterior part of the sternum, the xiphoid process and costal cartilages of adjacent ribs and attaches to the inner surface of the costal cartilages of the second to sixth ribs.
Nerve supply: Adjacent intercostal nerves.

LEVATORES COSTARUM

Attachments: They extend as eleven pairs of muscles from the transverse processes of C7 and T1–11 to the external surface of the ribs below their vertebral attachments.
Nerve supply: Dorsal rami of the corresponding thoracic nerves.

SERRATUS POSTERIOR — SUPERIOR

Attachments: Extends from the spinous processes of the two lower cervical and two upper thoracic vertebrae, to the outer side of the angles of the second to fifth ribs.
Nerve supply: By the appropriate intercostal nerves.

SERRATUS POSTERIOR — INFERIOR

Attachments: Extends from the spinous processes of the two lower thoracic and two upper lumbar vertebrae to the lower borders of the last four ribs.
Nerve supply: Ventral rami of the corresponding thoracic nerves.

For subcostales, transversus thoracis, levatores costarum, serratus posterior superior and inferior

Actions: They elevate or depress the ribs during respiration.

TESTING POSITION
Not applicable, as evaluation of strength requires specialized procedures.

Comments: These muscles are included for completeness.

Essential functions

Evaluating the muscles of micturition and defaecation

MUSCLES OF THE PELVIS
Levator ani — iliococcygeus
                — pubococcygeus
Coccygeus

MUSCLES OF THE PERINEUM
Transversus perinei superficialis
Transversus perinei profundus
Bulbospongiosus
Ischiocavernosus
Sphincter urethrae
Sphincter ani externus

Note: It is not appropriate to consider the surface markings and maximum extensibility of the muscles which make up the pelvic floor, therefore these headings have been omitted. As with muscles which contribute to other essential functions, the strength of those involved in micturition and defaecation cannot be assessed using the accepted procedures associated with manual muscle testing.

It is suggested that a Grade 5 be allocated when the patient can constrict the pelvic floor rapidly and a Grade 2 when there is an attempt to constrict the pelvic floor. Also, because of the nature of the assessment procedure these muscles are not included as part of the routine muscle test. Where there are indications of muscle weakness, however, the manual technique described here may be employed.

For a more detailed description of the methods of management of pelvic floor function the reader is referred to Shepherd & Montgomery (1983), Castleden, Duffin & Mitchell (1984) Gordon & Logue (1985).

LEVATOR ANI

Attachments: It is divided into two parts. Iliococcygeus extends from the spine of the ischium and the arch of the pelvic fascia to the coccyx and median raphe where it unites with its paired opposite. Pubococcygeus extends from the posterior surface of the pubis and the obturator fascia and unites with that of the opposite side in front of the coccyx.
Nerve supply: Pudendal S3, 4.

COCCYGEUS

Attachments: Extends from the spine of the ischium and the sacrospinous ligament to the coccyx and the fifth sacral vertebra.
Nerve supply: S3, 4.

Levator ani and coccygeus

Actions: These muscles constrict the rectum and vagina and form the muscular floor of the pelvis which supports the abdominal viscera when in the upright position.

TRANSVERSE PERINEI SUPERFICIALIS

Attachments: Extends from the ischial tuberosity and unites centrally with the muscle of the opposite side at the perineal body.
Nerve supply: Pudendal S2, 3, 4.

Actions: It probably stabilizes the peroneal body.

TRANSVERSE PERINEI PROFUNDUS

Attachments: Extends from the ramus of the ischium and the fascial sheath of the pudendal blood vessels to the perineal body.
Nerve supply: Pudendal S2, 3, 4.

Actions: It probably stabilizes the perineal body.

BULBOSPONGIOSUS

Attachments: Extends from the perineal body and the median raphe to the corpora cavernosa clitoridis in the female and to the dorsal surface of the penis in the male.
Nerve supply: S2, 3, 4.

Actions: In the female it constricts the vagina and in the male it aids urethral emptying, erection, and ejaculation.

ISCHIOCAVERNOSUS

Attachments: Extends from the tuberosity of the ischium to the crus clitoris in the female and to the crus penis in the male.
Nerve supply: S2, 3, 4.

Actions: In the female it compresses the crus clitoris and in the male the crus penis, contributing to the erection of both structures.

SPHINCTER URETHRAE

Attachments: Extends from the transverse perineal ligament and the adjacent fascia to the perineal body.
Nerve supply: Pudendal S2, 3, 4.

Actions: It constricts the urethra in both sexes.

For all of the above muscles

TESTING POSITION
Grade 5 (Not illustrated): In crook lying, with the patient suitably draped, two gloved lubricated fingers are inserted into the vagina. The patient constricts the vagina. (This is the same action as is used to stop the flow of urine in midstream.)
Grade 2 (Not illustrated): In crook lying, with the patient suitably draped, two gloved lubricated fingers are inserted into the vagina. The patient attempts to constrict the vagina.

Comments: This test is only applicable to female subjects.

SPHINCTER ANI EXTERNUS

Attachments: Encircles the lower two-thirds of the anal canal in a layered fashion.
Nerve supply: S2, 3, 4.

Actions: It constricts the anal canal.

TESTING POSITION
Grade 5 (Not illustrated): In side lying, with the hips in 90° of flexion and the patient suitably draped, one gloved lubricated finger is inserted into the rectum. The patient constricts the anus.
Grade 2 (Not illustrated): In side lying, with the hips in 90° of flexion and the patient suitably draped, one gloved lubricated finger is inserted into the rectum. The patient attempts to constrict the anus.

Comments: When any doubt about the function of the sphincter exists, observation is valuable. When functioning normally, the external sphincter maintains the anus in a retracted position between the gluteal folds. Where the sphincter muscle is weak or non functioning the anus appears more relaxed or even patulous and the rectal mucosa may be visible.

Alexander J, Molnar G E 1973 Muscular strength in children: Preliminary report on objective standards. Archives of Physical Medicine and Rehabilitation 54:424–427

Baldauf K L, Swenson D K, Medeiros J M, Radtka S A 1984 Clinical assessment of trunk flexor muscle strength in healthy girls 3 to 7 years of age. Physical Therapy 64(8): 1203–1208

Barrow H M, McGhee R 1979 A practical approach to measurement in physical education. Lea and Febiger, Philadelphia

Bogduk N 1980 A reappraisal of the anatomy of the human lumbar erector spinae. Journal of Anatomy 131(3):525–540

Bogduk N, Twomey L T 1987 Clinical anatomy of the lumbar spine. Churchill Livingstone, Melbourne

Bromley I 1985 Tetraplegia and paraplegia; a guide for physiotherapists, 3rd edn. Churchill Livingstone, Edinburgh

Castleden C M, Duffin H M, Mitchell E P 1984 The effect of physiotherapy on stress incontinence. Age and Ageing 13:235–237

Daniels L, Worthingham C 1986 Muscle testing, 6th edn. W B Saunders, Philadelphia

Evjenth O, Hamberg J 1985a Muscle stretching in manual therapy. A clinical manual, vol 1 Extremities. Alfta Rehab, Alfta

Evjenth O, Hamberg J 1985b Muscle stretching in manual therapy, A clinical manual, vol 11 Spinal column. Alfta Rehab, Alfta

Gosling J A, Harris P F, Humpherson J R, Whitmore I, Willan P L T 1985 Atlas of human anatomy. Churchill Livingstone, Edinburgh

Gordon H, Logue M 1985 Perineal muscle function after childbirth. Lancet 2:123–125

Gowitzke B A, Milner M 1980 Understanding the scientific basis of human movement. Williams and Wilkins, Baltimore

Green M, Moxham J 1983 Respiratory muscles. In Flenley D C, Petty T L (eds) Recent advances in respiratory medicine 3. Churchill Livingstone, Edinburgh

Griffin J W, McClure M H, Bertonini T E 1986 Sequential isokinetic and manual muscle testing in patients with neuromuscular disease. A pilot study. Physical Therapy 66(1):32–35

Janda V 1983 Muscle function testing. Butterworths, London

Kendall F P, McCreary E K 1983 Muscles, testing and function, 3rd edn. Williams and Wilkins, Baltimore

Lefkok M B 1986 Trunk flexion in healthy children aged 3 to 7 years. Physical Therapy 66(1):39–44

Leigh R T, Zee D S 1983 The neurology of eye movements. F A Davis, Philadelphia

McDonald C M, Jaffe K M, Shurtleff D B 1986 Assessment of muscle strength in children with meningomyelocele: accuracy and stability of measurements over time. Archives of Physical Medicine and Rehabilitation 67:855–861

Medical Research Council 1943 Aids to investigation of peripheral nerve injuries. HMSO, London

Miller A J 1982 Deglutition. Physiological Reviews 62(1):129–184

Molnar G E, Alexander J 1974 Development of quantitative standards for muscle strength in children. Archives of Physical Medicine and Rehabilitation 55:490–493

Pact V, Sirotkin-Roses M, Beatus J 1984 The muscle testing handbook. Little Brown and Co, Boston

Sanders M, Sanders B 1985 Mobility: active resistance training. In: Gould III J A, Davis G J (eds) Orthopedic and sports physical therapy. C V Mosby, St Louis

Shepherd A M, Montgomery E 1983 Treatment of genuine stress incontinence with a new perineometer. Physiotherapy 69:113

Twomey L T, Taylor J R 1987 Physical therapy of the low back. Churchill Livingstone, New York

Williams P L, Warwick R 1980 Gray's anatomy, 36th edn. Churchill Livingstone, Edinburgh

Wolff P H 1982 Theoretical issues in the development of motor skills. In: Lewis M, Taft L (eds) Development disabilities, theory, assessment and intervention. MTP Press, Lancaster

MUSCLE GRADING CHART: Trunk, upper limb, lower limb

CURTIN
University of Technology
Perth Western Australia
School of Physiotherapy
© 1988

Surname .... Forename ....

Address/Unit no ....

Sex .... Date of birth .... Occupation .... Handedness ....

Physician .... Tester ....

Diagnosis ....

....

....

....

....

....

Accessory descriptors

P   – pain
T   – abnormal muscle tone
ROM – altered range of movement
C   – contracture
PS – decreased proximal stability
F   – limited co-operation
OM – omitted

Muscle grades

0 – no function
1 – a flicker
1+ – some joint movement with gravity eliminated
2 – full ROM with gravity eliminated
2+ – full ROM with gravity eliminated + resistance
3 – full ROM against gravity
4 – full ROM against gravity + resistance
5 – normal

Bracketed muscles are tested together, thus all muscles in a group are allocated the same grade.

Trunk

NERVE	MUSCLE (innervation)	GRADE-LEFT (DATE)						GRADE-RIGHT (DATE)					
Accessory (XIth Cranial)	Sternocleidomastoid												
	Neck flexors (C1–6)												
	Neck extensors (C1–8, T1)												
	Neck side flexors (C3–8)												
	Neck rotators (C5–8)												
	Trunk extensors (T1–S3)												
	Rectus abdominis (T5–12)												
	Obliquus externus abdominis (T7–12)												
	Obliquus internus abdominis (T7–L1)												
Lumbar plexus	Quadratus lumborum (T12–L4)												
Sacral plexus	Pelvic floor muscles (S2–4)												
	Sphincter ani externus (S2–4)												

Upper limb

NERVE	MUSCLE (innervation)	GRADE-LEFT (DATE)						GRADE-RIGHT (DATE)					
Accessory	Trapezius												
	upper fibres (C3, 4)												
	middle fibres (C3, 4)												
	lower fibres (C3, 4)												
Dorsal scapular	Levator scapulae (C3–5)												
	Rhomboideus-major (C4, 5)												
	Rhomboideus-minor (C4, 5)												
Long thoracic	Serratus anterior (C5–7)												

Appendix 1

Upper limb (contd)

GRADE-LEFT (DATE)	NERVE	MUSCLE (innervation)	GRADE-RIGHT (DATE)
	Thoracodorsal	Latissimus dorsi (C6–8)	
	Upper & lower subscapular	Subscapularis (C5, 6)	
	Lower subscapular	Teres major (C6, 7)	
	Lat. & Med. pectoral	Pectoralis major (C5–T1)	
	Med. pectoral	Pectoralis minor (C7–T1)	
	Axillary	Deltoid: anterior fibres (C5, 6)	
		middle fibres (C5, 6)	
		posterior fibres (C5, 6)	
		Teres minor (C5, 6)	
	Suprascapular	Infraspinatus (C5, 6)	
		Supraspinatus (C5, 6)	
	Musculocutaneous	Coracobrachialis (C5–7)	
		Biceps brachii (C5, 6)	
	Musculocutaneous & Radial	Brachialis (C5–7)	
	Radial	Brachioradialis (C5, 6)	
		Triceps (C6–8)	
		Anconeus (C7, 8)	
		Ext. carpi radialis longus (C6, 7)	
		Ext. carpi radialis brevis (C7, 8)	
	Post. interosseus	Supinator (C5, 6)	
		Ext. digitorum (C7, 8)	
		Ext. digiti minimi (C7, 8)	
		Ext. indicis (C7, 8)	
		Ext. carpi ulnaris (C7, 8)	

Upper limb (contd)

GRADE-LEFT (DATE)					NERVE	MUSCLE (innervation)	GRADE-RIGHT (DATE)				
					Median	Ext. pollicis brevis (C7, 8)					
						Ext. pollicis longus (C7, 8)					
						Abd. pollicis longus (C7, 8)					
						Abd. pollicis brevis (C8, T1)					
						Opponens pollicis (C8, T1)					
						Palmaris longus (C7, 8)					
						Flex. carpi radialis (C6, 7)					
						Flex. digitorum superficialis (C7–T1)					
						Flex. pollicis brevis (C6–T1)					
						Lumbricals 1 & 2 (C8, T1)					
					Ant. interosseus	Pronator teres (C6, 7)					
						Pronator quadratus (C7–T1)					
						Flex. pollicis longus (C8, T1)					
						Flex. digitorum profundus 1 & 2 (C8, T1)					
					Ulnar	Lumbricals 3 & 4 (C8, T1)					
						Flex. carpi ulnaris (C7, 8)					
						Flex. digitorum profundus 3 & 4 (C8, T1)					
						Abductor digiti minimi (C8, T1)					
						Opponens digiti minimi (C8, T1)					
						Flex. digiti minimi (C8, T1)					
						Dorsal Interossei (C8, T1)					
						Palmar interossei (C8, T1)					
						Add. pollicis (C8, T1)					

Appendix 1

Lower limb

NERVE	MUSCLE (innervation)	GRADE-LEFT (DATE)						GRADE-RIGHT (DATE)					
Lumbar plexus	Psoas Major (L1–3)												
Femoral	Iliacus (L2, 3)												
	Sartorius (L2, 3)												
	Quadriceps femoris (L2–4)												
	Pectineus (L2, 3)												
Obturator	Adductor magnus (L2–4)												
	Adductor longus (L2–4)												
	Adductor brevis (L2–4)												
	Gracilis (L2, 3)												
	Obturator externus (L3, 4)												
Sacral plexus	Piriformis (L5–S2)												
	Gemellus superior (L5, S1)												
	Gemellus inferior (L5, S1)												
	Obturator internus (L5, S1)												
	Quadratus femoris (L5, S1)												
Superior gluteal	Tensor fascia latae (L4, 5)												
	Gluteus minimus (L5, S1)												
	Gluteus medius (L5, S1)												
Inferior gluteal	Gluteus maximus (L5–S2)												
Sciatic	Biceps femoris (L5–S2)												
Tibial branch	Semitendinosus (L5–S2)												
	Semimembranosus (L5–S2)												
	Popliteus (L4–S1)												

Lower limb (contd)

NERVE	MUSCLE (innervation)	GRADE-RIGHT (DATE)						GRADE-LEFT (DATE)					
Deep peroneal	Tibialis anterior (L4, 5)												
	Peroneus tertius (L5, S1)												
	Ext. hallucis longus (L5, S1)												
	Ext. hallucis brevis (S1, 2)												
	Ext. digitorum longus (L5, S1)												
	Ext. digitorum brevis (S1, 2)												
Superficial peroneal	Peroneus longus (L5–S2)												
	Peroneus brevis (L5–S2)												
Tibial	Tibialis posterior (L4, 5)												
	Gastrocnemius (S1, 2)												
	Plantaris (S1, 2)												
	Soleus (S1, 2)												
	Flex. hallucis longus (S2, 3)												
	Flex. digitorum longus (S2, 3)												
Medial plantar	Flex. digitorum brevis (S2, 3)												
Lateral plantar	Flex. digitorum accessorius (S2, 3)												
	Add. hallucis (S2, 3)												
	Abd. digiti minimi (S2, 3)												
	Dorsal interossei (S2, 3)												
	Plantar interossei (S2, 3)												
	Lumbricals 2–4 (S2, 3)												
Medial plantar	Lumbrical 1 (S2, 3)												
	Abductor hallucis (S2, 3)												
	Flex. hallucis brevis (S2, 3)												

COMMENTS:

Trunk:

Upper Limb:

Lower Limb:

CURTIN
University of Technology
Perth Western Australia
School of Physiotherapy
© 1988

MUSCLE GRADING CHART: Face and essential functions

Surname .... Forename ....

Address/Unit no ....

Sex ....Date of birth ....Occupation ....Handedness ....

Physician .... Tester ....

Diagnosis ....

....

....

....

....

....

Accessory descriptors

P — pain
T — abnormal muscle tone
ROM — altered range of movement
C — contracture
PS — decreased proximal stability
F — limited co-operation
OM — omitted

Muscle grades

0 — no function
2 — muscle activity less than normal
5 — normal

Bracketed muscles are tested together, thus all muscles in a group are allocated the same grade.

Swallowing

NERVE		MUSCLE	DATE						GRADE
Facial (VIIth Cranial)	Stage 1	Orbicularis oris							
		Buccinator							
Hypoglossal (XIIth Cranial)		Tongue – intrinsic							
		Tongue – extrinsic							
		Genioglossus							
		Hyoglossus							
		Chondroglossus							
		Styloglossus							
		Palatoglossus							
Vagus (Xth Cranial)									
Trigeminal & Facial (Vth & VIIth Cranial)		Suprahyoids							
	Stage 2	Pharyngeal constrictors							
	Stage 3	Infrahyoids							

COMMENTS:

Swallowing:

Face & Eyes

NERVE	MUSCLE (innervation)	GRADE-LEFT DATE					GRADE-RIGHT DATE				
Oculomotor (IIIrd Cranial)	Levator palpebrae superioris										
	Obliquus oculi inferior										
	Rectus superior										
	Rectus inferior										
	Rectus medialis										
Trochlear (IVth Cranial)	Obliquus oculi superior										
Abducent (VIth Cranial)	Rectus lateralis										
Trigeminal (Vth Cranial)	Masseter										
	Temporalis										
	Medial pterygoid										
	Lateral pterygoid										
Facial (VIIth Cranial)	Epicranius (frontal)										
	Corrugator supercilii										
	Obicularis oculi										
	Nasalis—Alar portion										
	Nasalis – transverse portion										
	Depressor septi										
	Procerus										
	Levator anguli oris										
	Levator labii superioris										
	Levator labii superioris alaeque nasi										
	Orbicularis oris										
	Risorius										

Appendix 2

Zygomaticus major	Zygomaticus minor	Mentalis	Buccinator	Depressor anguli oris	Platysma	Depressor labii inferioris

COMMENTS:

Face & Eyes:

166

GENERAL INDEX

ankle joint, muscles tested at, 96–102
central nervous system dysfunction, 15
    see also spinal cord lesions
cervical spine, muscle tests, 118–121
children, 13–14
confused or disorientated patients, 13
elbow joint, muscles tested at, 48–52
elderly patients, 13
essential functions, muscle tests,
    143–153
eye, muscle tests, 140–142
face, muscle tests, 129–142
finger joints, muscles tested at, 61–72
grading system, 7–8
    accessory descriptors, 8
hip joint, muscles tested at, 80–90
intellectually handicapped patients, 14
key to symbols, 29
knee joint, muscles tested at, 91–95
lower extremity, muscle tests, 80–117
lumbar spine, muscle tests, 122–128
manual muscle testing
    charting of results, 17–18
        see also Appendices
    grading system
    limitations, 4
    modifications, 12–14
    preparation, 11–12
    sequence, 9, 11–12
    special problems, 15
    weight bearing tests, 4, 114–115
mastication, muscle tests, 129
micturition and defaecation, muscle
    tests, 152–153
muscle contraction, 3–4, 9–10
muscular dystrophy, 14
neck and trunk, muscle tests,
    118–128
neonates, 14
nerve lesions, 15–16
obesity, 12
pain, 15
perceptual dysfunction, 13
respiration, muscle tests, 150–151
sacral spine, muscle tests, 122–128
scapula, muscle-tests, 30–37
sensory loss, 12
shoulder joint, muscle tests, 38–47
special groups of patients, 12–14
    children, 13–14
    confused or disorientated, 13
    elderly, 13
    intellectually handicapped, 14

neonates, 14
    obese, 12
spina bifida, 14
spinal cord lesions, 17
    dysfunction due to, 24–27
superior and inferior radio-ulnar joints,
    muscles tested at, 53–55
swallowing, muscle tests, 143–149
    stage 1, 144–147
    stage 2, 148
    stage 3, 149
thoracic spine, muscle tests, 122–128
thumb joints, muscles tested at, 73–79
toddlers' tests, 14, 23
toe joints, muscles tested at, 103–112
Trendelenberg sign, 114
upper extremity, muscle tests, 30–79
weight bearing, 4
    test, 114–115
wrist joint, muscles tested at, 56–60

MUSCLES TESTED

abdominal muscles *see* obliquus externus
    abdominis; obliquus internus
    abdominis; rectus abdominis;
    transversus abdominis
abductor digiti minimi
    foot, 110
    hand, 66
abductor hallucis, 109
abductor pollicis brevis, 77
abductor pollicis longus, 77
adductor brevis, 85
adductor hallucis, 111
adductor longus, 85
adductor magnus, 85
adductor pollicis, 78
anconeous, 52
aryepiglottic, 148
arytenoid
    oblique, 148
    transverse, 148
back extensors, 123
biceps brachii, 49
biceps femoris, 92
brachialis, 50
brachioradialis, 51
buccinator, 138, 144
bulbospongiosus, 153
chondroglossus, 145
coracobrachialis, 39
corrugator supercilii, 132
deltoid, 40
    anterior fibres, 40

    middle fibres, 43
    posterior fibres, 42
depressor anguli oris, 139
depressor labii inferioris, 139
depressor septi, 134
diaphragm, 151
digastric, 146
dorsal interossei
    foot, 110
    hand, 67
epicranius (frontal bellies), 132
extensor carpi radialis brevis, 59
extensor carpi radialis longus, 59
extensor carpi ulnaris, 60
extensor digiti minimi, 71
extensor digitorum, 69
extensor digitorum brevis, 106
extensor digitorum longus, 106
extensor hallucis brevis, 105
extensor hallucis longus, 104
extensor indicis, 70
extensor pollicis brevis, 75
extensor pollicis longus, 76
flexor carpi radialis, 58
flexor carpi ulnaris, 58
flexor digiti minimi, 63
flexor digitorum accessorius, 108
flexor digitorum brevis, 108
flexor digitorum longus, 108
flexor digitorum profundus, 65
flexor digitorum superficialis, 64
flexor hallucis brevis, 107
flexor hallucis longus, 107
flexor pollicis brevis, 74
flexor pollicis longus, 74
gastrocnemius, 98, 115
gemellus inferior, 89
gemellus superior, 89
genioglossus, 145
geniohyoid, 146
gluteus maximus, 81
gluteus medius, 87, 114
gluteus minimus, 88
gracilis, 85
hamstring group, 92–93
hyoglossus, 145
iliacus, 82
iliococcygeus *see* levator ani
iliocostalis cervicis, 117
iliocostalis lumborum, 123
iliocostalis thoracis, 123
iliopsoas, 82
inferior oblique (obliquus oculi inferior),
    141
inferior pharyngeal constrictor, 149
infrahyoids, 149

Index

infraspinatus, 46
intercostales, 151
interossei
  foot
    dorsal, 110
    plantar, 111
  hand
    dorsal, 67
    palmar, 68
interspinales, 117, 123
intertransversarii, 117, 123
ischiocavernosus, 153

lateral pterygoid, 131
latissimus dorsi, 41
levator anguli oris, 135
levator ani, 153
levatores costarum, 151
levator labii superioris, 136
  alaeque nasi, 136
levator palpebrae superioris, 133
levator scapulae, 31
levator veli palatini, 148
longissimus capitis, 117
longissimus cervicis, 117
longissimus lumborum, 123
longissimus thoracis, 123
longus capitis, 119
longus colli, 119
lumbricals
  foot, 112
  hand, 62

masseter, 130
medial pterygoid, 130
mentalis, 138
middle pharyngeal constrictor, 148
multifidus, 123
mylohyoid, 146

nasalis
  alar portion, 134
  transverse portion, 134
neck extensors, 117–118
neck flexors, 119–120
neck side flexors, 119–121

oblique arytenoid, 148
obliquus capitis inferior, 117
obliquus capitis superior, 117
obliquus externus abdominis, 126
obliquus internus abdominis, 126
obliquus oculi inferior, 141
obliquus oculi superior, 141
obturator externus, 89
obturator internus, 89
omohyoid, 149
opponens digiti minimi, 72

opponens pollicis, 79

palatoglossus, 146
palatopharyngeus, 148
palmar interossei, 68
palmaris longus, 57
pectineus, 85
pectoralis major, 45
pectoralis minor, 37
pelvic floor muscles, 153
peroneus brevis, 102
peroneus longus, 102
peroneus tertius, 100
pharyngeal constrictors, 148, 149
piriformis, 89
plantar interossei, 111
plantaris, 98
platysma, 139
popliteus, 95
procerus, 135
pronator quadratus, 55
pronator teres, 55
psoas major, 82
pterygoid
  lateral, 130
  medial, 131
pubococcygeus see levator ani

quadratus femoris, 89
quadratus lumborum, 128
quadriceps femoris, 94, 115

rectus abdominis, 125
rectus capitis anterior, 119
rectus capitis lateralis, 121
rectus capitis posterior major, 117
rectus capitis posterior minor, 117
rectus femoris, 94
rectus muscles of the eye, 141
rhomboideus major, 34
rhomboideus minor, 34
risorius, 137
rotatores, 123
  cervical, 117

salpingopharyngeus, 148
sartorius, 84
scalenus anterior, 121
scalenus medius, 121
scalenus posterior, 121
semimembranosis, 92
semispinalis capitis, 117
semispinalis cervicis, 117
semispinalis thoracis, 123
semitendinosis, 92
serratus anterior, 36
serratus posterior

inferior, 151
superior, 151
soleus, 97, 115
sphincter ani externus, 153
spinalis capitis, 117
spinalis cervicis, 117
spinalis thoracis, 123
splenius capitis, 117
splenius cervicis, 117
sternocleidomastoid, 119
sternocostalis, 151
sternohyoid, 149
sternothyroid, 149
styloglossus, 145
stylohyoid, 149
stylopharyngeus, 148
subclavius, 37
subcostales, 151
subscapularis, 47
superior oblique see obliquus occuli
  superior
superior pharyngeal constrictor, 148
supinator, 54
suprahyoids, 146
supraspinatus, 44

temporalis, 130
tensor fasciae latae, 83
tensor veli palatini, 148
teres major, 47
teres minor, 46
thyroarytenoid, 148
thyroepiglottic, 148
thyrohyoid, 149
tibialis anterior, 99
tibialis posterior, 101
tongue
  extrinsic muscles, 145
  intrinsic muscles, 144
transverse arytenoid, 148
transverse perinei profundus, 153
transverse perinei superficialis, 153
transverse thoracis, 151
transversus abdominis, 125
trapezius
  lower fibres, 35
  middle fibres, 33
  upper fibres, 32
triceps, 52

vastus intermedius, 94
vastus lateralis, 94
vastus medialis, 94

zygomaticus major, 137
zygomaticus minor, 137